Introductory Statistics

Authored By

Alandra Kahl

*Department of Environmental Engineering, Penn State
Greater Allegheny, McKeesport, Pennsylvania
USA*

Introductory Statistics

Author: Alandra Kahl

ISBN (Online): 978-981-5123-13-5

ISBN (Print): 978-981-5123-14-2

ISBN (Paperback): 978-981-5123-15-9

Published by Bentham Science Publishers Pte. Ltd. Singapore. All Rights Reserved.

First published in 2023.

need for a court order if at any point you breach any terms of this License Agreement. In no event will any delay or failure by Bentham Science Publishers in enforcing your compliance with this License Agreement constitute a waiver of any of its rights.

3. You acknowledge that you have read this License Agreement, and agree to be bound by its terms and conditions. To the extent that any other terms and conditions presented on any website of Bentham Science Publishers conflict with, or are inconsistent with, the terms and conditions set out in this License Agreement, you acknowledge that the terms and conditions set out in this License Agreement shall prevail.

Bentham Science Publishers Pte. Ltd.
80 Robinson Road #02-00
Singapore 068898
Singapore
Email: subscriptions@benthamscience.net

CONTENTS

PREFACE

Statistics is a complex and multi-faceted field that is relevant to many disciplines including business, science, technology, engineering and mathematics. Statistical analysis and research is critical to understanding data sets, compiling and analyzing scientific results and presenting findings. Without statistics, research would grind to a halt for lack of support and discourse regarding presentation of results. We rely on statistics and analysis to make sense of patterns, nuances and trends in all aspects of science.

This volume presents a brief but thorough overview of common statistical measurements, techniques and aspects. It discusses methods as well as areas of presentation and discourse. Chapter 1 presents an introduction to the field and relevant data types and sample data. Chapter 2 highlights summarizing and graphing, including relevant charts such as histograms, box plots, and pie charts. Chapter 3 discusses the basic concepts of probability by discourse on sample events, sample spaces, intersections, unions and complements. Chapter 3 also encompasses conditional probability and independent events as well as basic principles and rules. Chapter 4 targets random variables, including discrete values and binomial distributions. Chapter 5 summarizes continuous random variables as well as the normal distribution. Chapter 6 surveys sampling distributions, the sample mean and the central limit theorem. Chapter 7 holds forth on estimation, including intervals of confidence and the margin of error. Chapter 8 covers hypothesis testing as well as the t-test and z- test. Chapter 9 speaks about the important topics of correlation and regression. Chapter 10 briefly examines the ethics associated with statistics, including the tenets of ethical conduct for those in the discipline.

In short, this book presents a brief scholarly introduction to the chief topics of interest in statistics. It is hoped that this volume will provide a better understanding and reference for those interested in the field as well as the greater scientific community.

I am grateful for the timely efforts of the editorial personnel, particularly Mrs. Humaira Hashmi (Editorial Manager Publications) and Mrs. Fariya Zulfiqar (Manager Publications).

CONSENT FOR PUBLICATION

Not applicable.

CONFLICT OF INTEREST

The author declares no conflict of interest, financial or otherwise.

ACKNOWLEDGEMENT

Declared none.

Alandra Kahl
Department of Environmental Engineering
Penn State Greater Allegheny
McKeesport, Pennsylvania
USA

CHAPTER 1

Introduction to Statistics

Abstract: The field of statistics is vast and utilized by professionals in many disciplines. Statistics has a place in science, technology, engineering, medicine, psychology and many other fields. Results from statistical analysis underlying both scientific and heuristic reasoning, and therefore, it is important for everyone to grasp basic statistical methods and operations. A brief overview of common statistical methods and analytical techniques is provided herein to be used as a reference and reminder material for professionals in a broad array of disciplines.

Keywords: Analysis, Heuristic reasoning, Scientific reasoning, Statistical methods.

INTRODUCTION

The field of statistics deals with the collection, presentation, analysis and use of data to make decisions and solve problems. Statistics is important for decision-making, cost-benefit analysis and many other fields. A good grasp of statistics and statistical methods can be beneficial to both practicing engineering as well as practicing businessmen. Specifically, statistical techniques can be a powerful aid in designing new products and systems, improving existing designs and developing and improving production processes. Statistical methods are used to help decide and understand variability. Any phenomenon or operation which does not produce the same result every time experiences variability. Individuals encounter variability in their everyday lives, and statistical thinking and methods can be a valuable aid to interpret and utilize variability for human benefit. For example, consider the gas mileage of the average consumer vehicle. Drivers encounter variability in their gas mileage driven by the routes they take, the type of gas they put in their gas tanks, and the performance of the car itself as examples. There are many more areas in which variability is introduced, all of which drive variability related to the gas mileage of the individuals' car. Each of these are examples of potential sources of variability in the system of the car. Statistics gives us a framework for describing this variability as well as for learning which potential sources of variability are the most important or have the greatest impacts on performance. Statistics are numerical facts or figures that are observed or obtained from experimental data.

Data is typically collected in one of two ways, either observational study or designed experiments. Data can also be obtained *via* random sampling or randomized experiments, but it is difficult to discern whether the data has any statistical significance- that is, is the difference found in the sample strictly related to a specific factor [1]. Simply put, is there a cause-and-effect relationship between the observed phenomena and the result? It is far more useful to collect data using observational study or designed experiments for statistics, as researchers can better narrow, understand and discard confounding factors within the gathered data set.

The first way that data can be collected is by observational study. In an observational study, the researcher does not make any impact on the collection of the data to be used for statistics; rather, they are taking data from the process as it occurs and then trying to ascertain if there are specific trends or results within that data [1]. For example, imagine that the interested researcher was curious about whether high iron levels in the body were associated with an increased risk of heart attacks in men. They could look at the levels of iron and other minerals within a group of men over the course of five years and see if, in those individuals who displayed high iron levels, there were more heart attacks. By simply tracking the subjects over time, the researchers are performing an observational study [1]. It is difficult in an observational study to identify causality as the observed statistical difference could be due to factors other than those the researchers are interested in, such as stress or diet in our heart attack example. This is because the underlying factor or factors that may increase the risk of heart attack was not equalized by randomization or by controlling for other factors during the study period, such as smoking or cholesterol levels [2]. Another way that observational data is obtained to by data mining, or gleaning information from previously collected data such as historical data [1]. This type of observational study is particularly useful in engineering or manufacturing, where it is common to keep records on batches or processes. Observational engineering data can be used to improve efficiency or identify shortcomings within a process by allowing a researcher to track a trend over time and make conclusions about process variables that may have positively or negatively caused a change in the final product.

The second way that data can be obtained for statistical work is through a designed experiment. In a designed experiment, the researcher makes deliberate or purposeful changes in the controllable variables of a system, scenario or process, observes the resultant data following these changes and then makes an inference or conclusion about the observed changes. Referring to the heart attack study, the research could design an experiment in which healthy, non-smoking males were given an iron supplement or a placebo and then observe which group had more heart attacks during a five-year period. The design of the experiment now controls for underlying factors, such as smoking, allowing the researchers to make a stronger conclusion or inference about the obtained data set. Designed experiments play an important role in science, manufacturing, health studies and engineering as they help researchers eliminate confounding factors and come to strong conclusions [1]. Generally, when products, guidelines or processes are designed or developed with this framework, the resulting work has better performance, reliability and lower overall costs or impacts. An important part of the designed experiments framework is hypothesis testing. A hypothesis is an idea about a factor or process that a researcher would like to accept or reject based on data. This decision-making procedure about the hypothesis is called hypothesis testing. Hypothesis testing is one of the most useful ways to obtain data during a designed experiment, as it allows the researcher to articulate precisely the factors which the researcher would like to prove or disprove as part of the designed experiment [1].

Modelling also plays an important role in statistics. Researchers interested in statistics can use models to both interpret data as well as to construct data sets to answer hypotheses. One type of model is called a mechanistic model. Mechanistic models are built from underlying knowledge about physical mechanisms. For example, Ohm's law is a mechanistic model which relates current to voltage and resistance from knowledge of physics that relates those variables [1]. Another type of model is an empirical model. Empirical models rely on our knowledge of a phenomenon but are not specifically developed from theoretical or first principles understanding of the underlying mechanism [3]. As an example, to illustrate the difference between mechanistic models and empirical models, consider the bonding of a wire to a circuit board as part of a manufacturing process. As part of this process, data is collected about the length of the wire needed, the strength of the bond of the wire to the circuit and the amount of solder needed to bond the wire. If a researcher would like to model the amount of solder needed to bond the wire related to the amount of force required to break the bond, they would likely use an empirical model as there is no easily applied physical mechanism to describe this scenario. Rather, the researcher determines the relationship between the two factors by creating a plot that compares them. This type of empirical model is called a regression model [1]. By estimating the parameters in regression models, a researcher can determine where there is a link between the cause and effect of the observed phenomena.

Another type of designed experiment is factorial experiments. Factorial experiments are common in both engineering and biology as they are experiments in which several factors are varied together to study the joint effects of several factors. Returning to the circuit board manufacturing example, an interested researcher could vary the amount of solder along with the length of wire used to determine if there are several alternative routes to obtain the strongest connection for the wire to the circuit board. In factorial experimental design, as the number of factors increases, the number of trials for testing increases exponentially [1]. The amount of testing required from study with many factors could quickly become infeasible from the viewpoint of time and resources. Fortunately, where there are five or more factors, it is usually unnecessary to test all possible combinations of factors. In this instance, a researcher could use a fractional factorial experiment, which is a variation on the factorial experiment in which only a subset of the possible factor combinations is tested. These types of experiments are frequently used in industrial design and development to help determine the most efficient routes or processes.

DATA TYPES

There are many different types of data that are utilized in statistics. Data within statistics is also known as variables. We will discuss six different types of variables within this text: independent, dependent, discrete, continuous, qualitative and quantitative variables [2]. Variables, as a general definition, are the properties or characteristics of some event, object or person that can take on different values or amounts. In designed experiments and hypothesis testing, these values are manipulated by the researcher as part of the study. For example, in the heart attack study, the researcher might vary the amount of iron in the supplement an individual received as part of the variables within the study. That variable is then referred to as the independent variable. In the same study, the effect of this iron supplement change is measured on the prevalence of heart attacks. The increase or decrease of a heart attack related to the amount of iron received in the supplement is referred to as the dependent variable. In general, the variable that is manipulated by the researcher is the independent variable, and its effects on the dependent variable are measured [1]. An independent variable can also have levels. For example, if control is included in the heart attack study, where participants receive a set amount of iron in the supplement, then the experiment has two levels of independent variables. In general, the number of independent variable levels corresponds to the number of experimental conditions within the study [4]. An important distinction between variables is that of qualitative and quantitative variables. Qualitative variables are variables that are not expressed in a numerical fashion, for instance, the eye or hair color of an individual or their relative girth or shape [2]. For example, when describing a subject, a researcher might refer to a body type as a pear shape. This variable is a qualitative type of variable as it does not have a numerical association. Qualitative-type variables can also be called categorical variables. Quantitative variables are those variables that are associated with a numerical value. For example, the grams of iron received in a supplement with the heart attack study would be a quantitative type of variable. Variables can also be discrete or continuous [2]. Discrete variables are those variables that fall upon a scale or within a set range. A good example of a discrete variable is the age range of a selection of patients within the study of a researcher. For example, the desired range of participants may be males between the age of 35 and 50. The age of each participant within the study falls upon a discrete scale with a range of 35 to 50 years of age. Each year is a discrete step; when an individual reports their age, it is either 35 or 36, not 36.5. Other variables, such as time spent to respond to a question, might be different, for example, this type of answer could be anywhere from 3.57 to 10.8916272 seconds. There are no discrete steps associated with this type of data, therefore the data is described as continuous rather than discrete [2]. For datasets like this, it is often practical to restrict the data by truncating the value at a set point, for example, at the tens or thousandths place, so it is not truly a continuous set.

Sample Data

When dealing with statistical data, it is important to identify the difference between population data sets and sample data sets. The type of data set utilized is important to understand as it is relevant to the available statistical tests that can be performed using that data set. For example, a small data set may necessarily be excluded from a statistical test that requires more results, such as a standard deviation-type statistical test [5]. Population data refers to the entire list of possible data values and contains all members of a specified group [2, 3]. For example, the population of all people living in the United States. A sample data set contains a part, or a subset of a population. The size of a sample data set is always less than that of the population from which it is taken. For example, some people living in the United States. Another way to think about the difference between population data and sample data would be to consider the heart attack example from earlier in the chapter. In this example, for population data, one might consider the entire population of males within the United States between the ages of 35-50 who have experienced a heart attack. A sample data set from this population might be only males who were taking an iron supplement who had experienced a heart attack. When performing calculations related to sample data sets *versus* population data sets, statisticians use the large letter N for the number of entries of population data and the small letter n for the number of entries of sample data [2, 3]. When calculating the mean for both types of data sets, for population data, the term $\bar{x}$ is used, while the term μ is used for the calculation of the mean for sample data sets.

For sample data sets, it is important to remember that these data sets are only parts of a whole, therefore when data is chosen for sampling, it is important to be mindful of the demographics of the data [3]. For example, if a data set represents a population that is 60% female and 40% male, the sample data set should also reflect this demographic breakdown.

Sample data sets are particularly important in marketing [3]. For example, imagine a business wants to sell a product to a subset of its current customers who don't yet own that product. The marketing department makes up a leaflet that describes the aspects of the products, the advantages of owning the product in addition to the company's other offerings, *etc.* The business estimates that of their 1 million customers, about 8 percent of them will buy the product, or about 80,000. Does the company send out 1 million leaflets to attempt to capture the 80,000 interested customers? No, they will put together a sample data set of customers who are likely to be most interested, based on previous purchases, business demographics, age, income and so on. This sample set of customers will reflect a subset of the million existing customers, which will allow the marketing department to reach a wide audience without blanketing their entire customer base in leaflets.

Sample data sets are also utilized heavily in data modelling [6]. When building a data model, it is important to test and train the model using sample data sets for validation. At least two datasets are typically used, but models may use up to as many as ten datasets to complete and cross-validate their model and its representative tools or outputs [3]. The number of datasets utilized for validation depends on the amount of cross-validation desired for the data model. Cross-validation is the comparison of results from distinct datasets to validate precision [3]. If greater precision is required, more datasets are used to achieve greater cross-validation. The purpose of cross-validation is to minimize errors in analysis by using each record once for testing and once for training. The sets are then swapped, and the experiment is repeated. The total error is determined by summing up the total errors for both runs [3]. It is important to note that the data samples occur in equal-sized subsets in each dataset used in this way for cross-validation. For example, if there are ten records in sample dataset A for cross-validation, there must also be ten records in sample dataset B. This is called the k-fold cross-validation method, where k represents the number of partitions, runs, and times the method is repeated [5]. For a special case of the cross-validation method, the method sets k=N, where N is the size of the datasets [3]. This is a special case which requires many subsets and generates a high degree of validation, resulting in a well-trained and well-distributed data model. The extraction of datasets for cross-validation and modelling requires that the distribution of results for each variable will be the same in the sample dataset as in the larger dataset [3]. For example, if a 5% sample dataset is desired to be extracted from a data set containing 1 million entries, one cannot simply select the first 50,000 entries in the database, as these entries may be all women and not represent the distribution of 40% women and 60% men that are present in the entire dataset. To avoid this bias, a random data extraction method could be used to construct the sample dataset, such as the rand () function in Microsoft Excel. Another method of extracting random variables is to use the Java coding language to create a random floating-point variable which can then be pinned to extract a random set of variables in a large dataset. The Oracle brand software can also do this, using a random integer generating function to seed a dataset for sample extraction [3]. The random method of dataset construction also needs to be checked against the full dataset using statistical analysis to determine if the sample dataset represents the full dataset correctly. One method of determining this is to set an error tolerance margin for the sample dataset [6]. For example, an error tolerance margin of 10% would allow for variation from the representative distributions of the full dataset to be no greater than 10%. Statistics such as the numerical average, standard deviation, modal values, correlations and frequencies between the sample dataset and full dataset are utilized to be sure that the sample dataset is correctly represented with respect to the overall dataset [7]. It is also important to note that the method of random data selection to generate sample datasets also does not build in error tolerance, so the same data may be selected twice, or similar records may be included that do not well represent the demographic of the overall dataset. Therefore, it is important to perform additional statistical analysis on datasets before they are utilized and to interrogate datasets to make sure they are unique and non-repeating records or values [8]. Repeated records or values can also artificially inflate values from cross-validation sets as those sets will match more frequently and appear to increase the model validity, while they are not actually returning unique values.

It is also important when extracting multiple datasets to not repeat records. Records in each sample dataset should be exclusive and unique [3]. This is done by removing or flagging the values in sequential datasets in the main model, using a qualifier or director such as "not in" when choosing the data for the next sample dataset.

Validation of sample datasets is done in several ways. A simple way to check the validity and quality of data is to graph the data for the distribution of the key variables [7]. The data should ideally follow a uniform or Gaussian-shaped bell curve distribution. Outlying variables can then be inspected to eliminate extreme values from the ends of the curve.

CONCLUSION

In conclusion, the field of statistics has many methods of data collection. Careful collection of data and validation of data sets is vital to achieving correct and valuable answers to statistical questions. Sample data can be used as a baseline for data sets when performing necessary statistical modelling, or population data can be used to understand a large subset of data or a particular group. There are many choices when collecting data as well as utilizing data or data subsets, and it is important to make sure that methods are well documented and understood when beginning.

CHAPTER 2

Summarizing and Graphing

Abstract: Data captured during analysis can be easily summarized using visual techniques such as graphing. Graphing is utilized to both summarize and convey information in a clear and readable format. Graphing techniques discussed in this chapter include frequency distributions, histograms, box plots and Pareto charts.

Keywords: Box plot, Frequency distribution, Histogram.

INTRODUCTION

Visualization of data *via* graphical means is a common way to convey statistical information. Data is easily summarized by several graphical techniques, including frequency distributions, histograms, box plots, Pareto charts, and dot plots, among others. By showing data in graphical form, researchers can pinpoint trends, isolate outliers and approve or disprove hypotheses. Showing data in graphs allows for broader distribution of the dataset as well as simple interpretations of data. Differentiations of data can be accomplished by outline or color difference, or other charting techniques may be used. Graphing is important to convey information as well as support further data analysis.

FREQUENCY DISTRIBUTIONS AND HISTOGRAMS

Once data has been collected, the next step for data analysis is to organize the data to determine if there are meaningful trends or patterns present within the dataset. A common method for organizing data is to construct a frequency distribution or frequency table. This type of data organization is a tabulation of the number of individuals in each category organized graphically with respect to the scale of measurement [9]. Frequency distribution is useful because it allows a researcher to understand overarching trends in the dataset briefly. For example, by graphing data this way, it can be seen if observations are high or low, or concentrated around one area of the scale [10]. By showing how individual distributions relate to the entire dataset, trends can be seen.

Frequency distributions or frequency tables, are constructed to show the different categories of measurement within the dataset as well as the number of observations within each set. An example of a frequency table is shown in Fig. (1).

Frequency distribution of the resting pulse rate in healthy volunteers (N = 63)

Pulse/min	Frequency	Cumulative frequency	Relative cumulative frequency (%)
60–64	2	2	3.17
65–69	7	9	14.29
70–74	11	20	31.75
75–79	15	35	55.56
80–84	10	45	71.43
85–89	9	54	85.71
90–94	6	60	95.24
95–99	3	63	100

Fig. (1). Frequency table.

It is important to have an idea of the maximum and minimum ranges of values on the scale of the dataset before beginning to construct a frequency distribution so that the measurement scale is chosen appropriately. The selected data range is then divided into intervals called class intervals. These class intervals occur arbitrarily in order to minimize bias within the data sorting. It is also important to choose the correct amount of class intervals for the dataset. If there are too few class intervals, the dataset is too bulky, and it is difficult to see deviations. If there are too many class intervals, the dataset divisions are too small and small deviations are magnified. Generally, between six and fourteen class intervals are sufficient. It is also important to know the width of the class. The class width can be calculated by dividing the range of observations by the number of classes [9]. It is desirable to have equal class widths. Unequal class widths should only be used when there are large gaps in the data. Class intervals should also be mutually exclusive and nonoverlapping [11]. Therefore, data within one class is not repeated within another class. Finally, for greater precision and data sorting, open-ended classes at the lower and high ends of the data set should be avoided. For example, classes <10 or >100.

There are many ways to represent frequency data graphically. A commonly used way to show frequency distribution is called a histogram. A histogram is shown in Fig. (**2**).

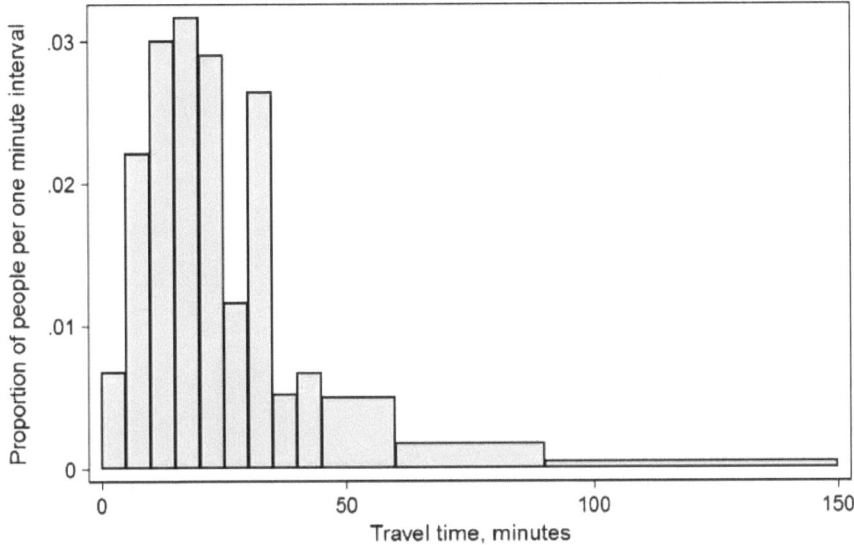

Fig. (2). A histogram [Source: WFP / Wikimedia Commons / CC BY 3.0].

A histogram shows the variable of interest in the x-axis and the frequency of the occurrence of that variable or the number of observations in the y-axis. Percentages can also be used if the objective is to compare two histograms having a different number of subjects. A histogram is used to depict the frequency when data are measured on an interval or ratio scale [10]. While a histogram may appear at first glance to look like a bar chart, there are three major differences between these graphical representations of data. In a histogram, there is no gap between the bars as the variable is continuous. Only if there is a large gap in the data scale will a gap occur in the histogram. In a histogram, the width of the bars is dependent on the class interval. If there are unequal class intervals, the widths of the bars of the histogram will reflect this inequality. Thirdly, in a histogram, the area of each bar corresponds to the frequency, where in a bar chart, it is the height that shows the frequency [10].

Histograms can also be used to generate other representations of the dataset. For example, a frequency polygon can be constructed by connecting the midpoints of the tops of the bars of a histogram using straight lines. A frequency polygon is shown in Fig. (**3**).

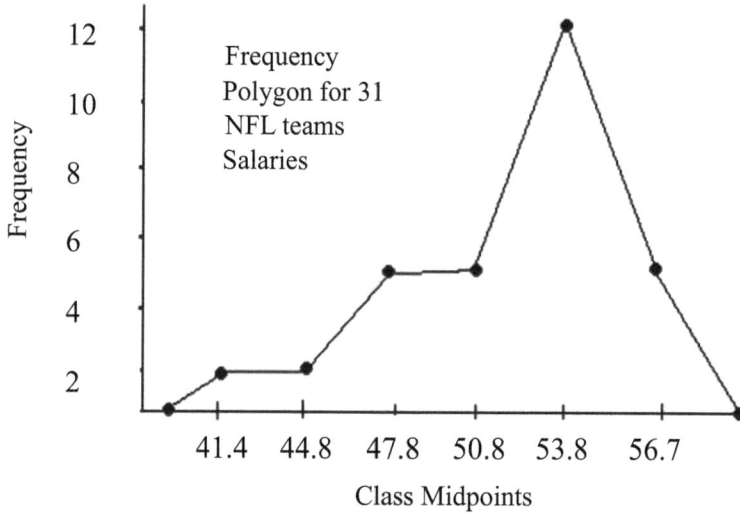

Fig. (3). Frequency polygon [Source: Wikimedia Commons].

This looks like a trendline and aids in the easy comparison of two frequency distributions. When the class intervals are narrow, and the two frequency is large, the line will smooth and result in a frequency curve.

Box and whisker plots can also be used to show the distribution of data. (Fig. **4**) shows an example of a box and whisker plot.

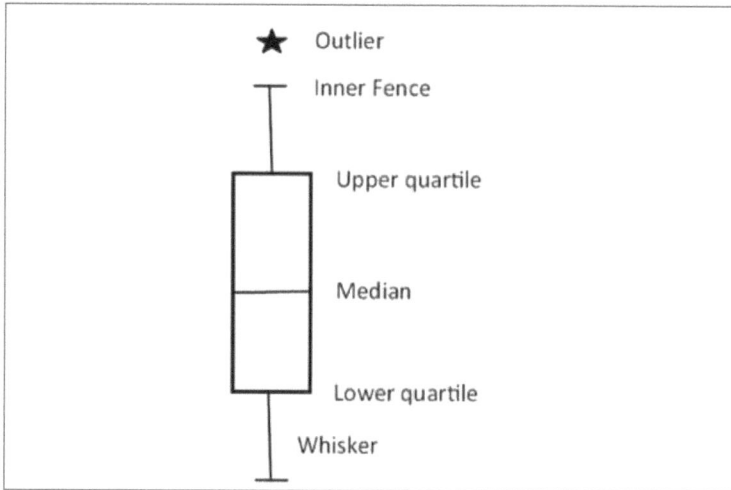

Fig. (4). Box and whisker plot [Source: Google Images].

This type of graphical data representation originated in 1977. The box in the box and whisker plot represents the middle 50% of the data set, with the ends of the box corresponding to the upper and lower quartiles of the data set (75% and 25%, respectively) [10]. The length of the box corresponds to the variability of the data, while the line inside of the box represents the median. If the line is closer to the bottom of the box, the data are skewed with respect to the lower quartile and similarly, if the line is closer to the top of the box, the data are skewed to the upper quartile [12]. The whiskers that extend outside of the top and bottom of the box represent the interquartile range and are 1.5 times the length of the box [12]. The end of the whisker is called the inner fence, and any variable outside of it is an outlier. If the distribution is symmetrical, then the whiskers will be of equal length. If the distribution is asymmetrical, the whisker will be shorter on the lesser side of the box.

GRAPHS

Graphing is particularly important in statistics to convey complex information from large groups of data or complex situations *via* visual models. These complex numerical stories need easy-to-interpret and understand visual representations, which is why graphing and choosing the correct graph for the statistical dataset is so important. A good graph conveys information quickly and easily to the user, gives a way to compare features of the data and shows the salient features of the numerical analysis concisely. There are seven different types of graphs that are commonly used in statistics: the Pareto diagram or bar graph, the pie chart or circle graph, the histogram, the stem and leaf plot, the dot plot, the scatterplot and the time series graph. Each type of graph has a different purpose and conveys data in a different way. We will now explore each type of graph, their uses, and for what type of data it is most used.

The Pareto diagram or bar graph is a very common way to represent and compare data in a small data set [11]. A simple bar chart allows the comparison of different data using bars or lengths. Data can be grouped either horizontally or vertically, allowing for easy comparison of data within a given set. Bar charts are particularly useful for qualitative data but can also be used for quantitative data to show the greatest or least value or accumulation in a set quickly [11]. The independent variable is grouped on the x-axis, while the dependent variable is grouped on the y-axis. Bar graphs can either be single, grouped or stacked bars. For a single bar chart, only one type of data is represented [13]. Bars are made of variable lengths to represent the magnitude of that variable for a simple and easy comparison of the single variable of interest. Fig. (5) shows an example of a single bar chart.

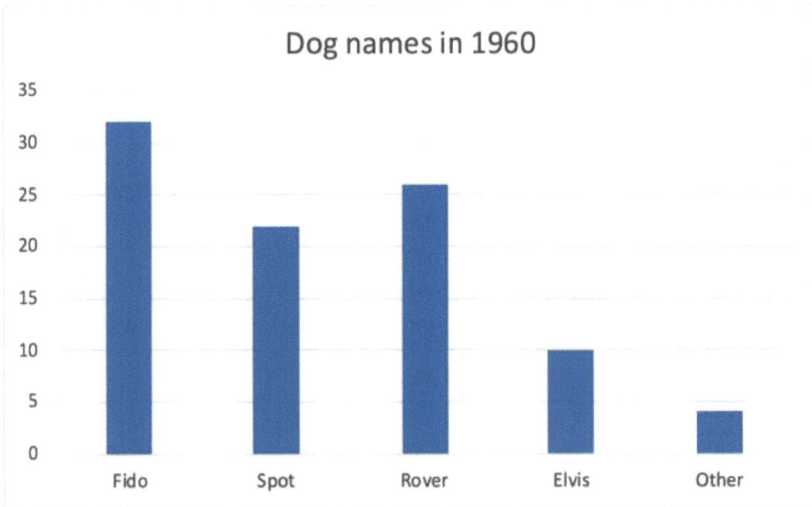

Fig. (5). Single bar chart showing the frequency of dog names in 1960. [Created by author].

A grouped bar chart is a type of bar chart that is used to represent values for more than one item that shares the same category. Grouped bar charts are used to show additional complexity within a dataset, for example, by adding another factor to the existing dataset or showing another aspect of the dataset [6, 13]. In the example shown in Fig. (**6**), the scores on an exam are given by the magnitude of score as well as broken out by gender.

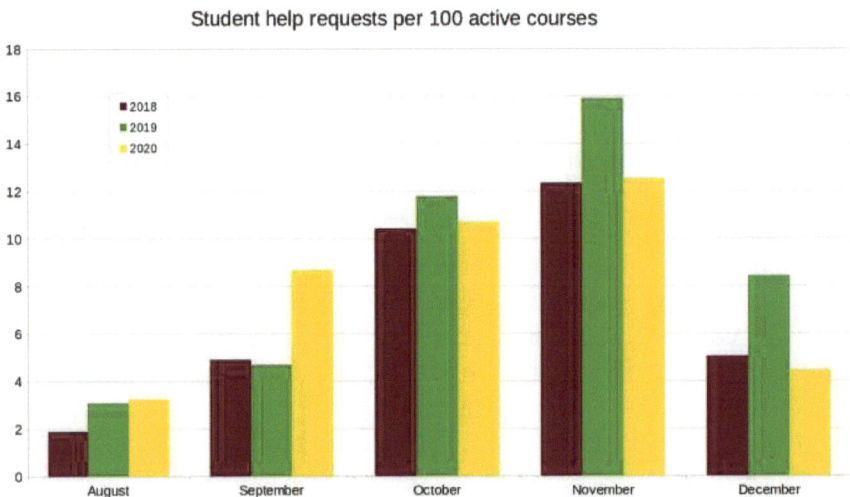

Fig. (6). Graph of the year-by-year rate of help requests from student editors in the Wikipedia Student Program Each Fall semester [Source: Wikimedia Commons].

There are also stacked bar charts. These types of bar charts show a further breakdown of data into aggregate factors by breaking bars into subsets within a group. An example of a stacked bar chart is shown in Fig. **(7)**.

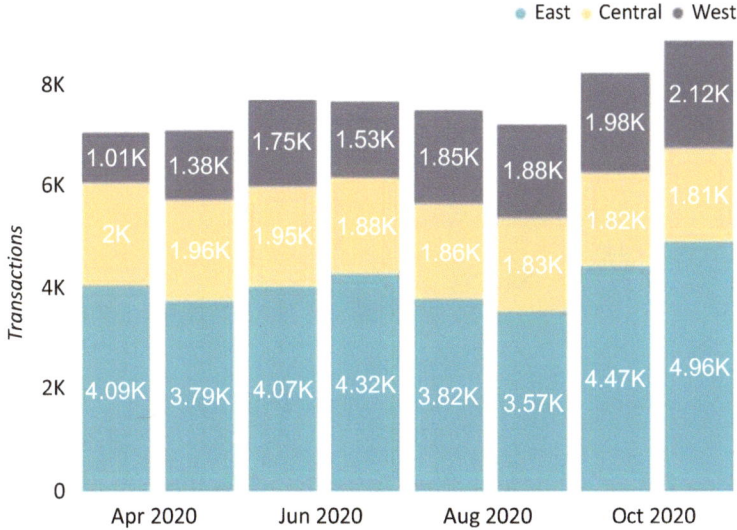

Fig. (7). Stacked bar chart showing revenue for a retailer in various areas.

There is also a type of bar chart called the Pareto chart. The Pareto chart or graph was developed by Vilfredo Pareto in 1909 [11]. A Pareto chart contains both bars and a line graph to show individual values in descending order using the bars and a line to show the cumulative total of those values. As there are two visual representations on a Pareto chart, there are two axes present. The left axis is the axis used for the bars and usually shows the frequency of the occurrence of the factor or other unit of measure, while the right axis shows the cumulative percentage of the number of occurrences or another factor on the bar chart [11]. In a Pareto chart, the value represented by the bar is always plotted in descending order, while the line shows each percentage weight of each of the factors on the bar chart, counting to 100% or the total of the values [14]. Pareto sought to use his chart to show economic disparities by plotting income *versus* population size in his original example, but Pareto chart can be used to show a wide variety of data, including the most frequent type of defect in a product, most common customer complaint, or the most ordered dish in a restaurant. The purpose of a Pareto chart is to highlight the most important item in a large set of factors. (Fig. **8**) shows an example of a Pareto chart showing the most frequent reasons for employee lateness to work.

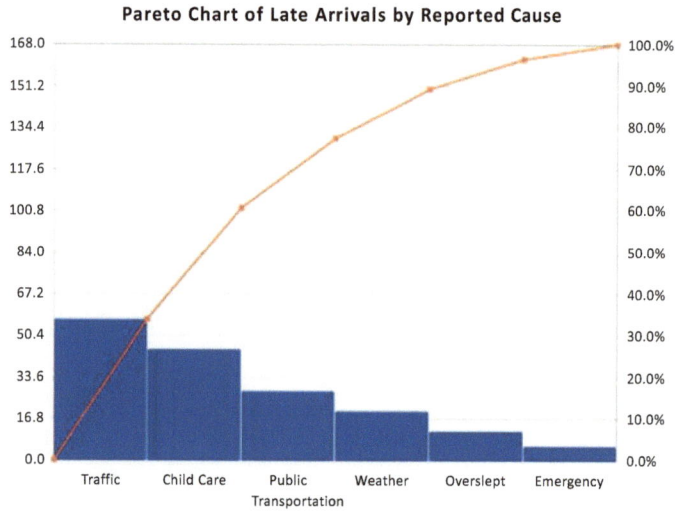

Fig. (8). Pareto chart example [Source: Wikimedia Commons].

Another type of chart commonly used in statistics is the pie chart or circle graph. The pie chart or circle graph is helpful for the presentation of qualitative data, where the information displayed is shown in relation to the whole. This is particularly useful for traits or characteristics, or data is then broken down into percentages. Each slice or portion of the pie chart corresponds to a categorical variable, and the size of each slice in area corresponds to the portion of the whole the categorical variable represents [12]. Pie charts are best used to show the contribution of parts to a whole, such as votes during an election or the proportion of fruits, vegetables and grains to an overall diet. Data shown in a pie chart is often paired with a data table to provide more details about the data [12]. For example, in the pie chart shown in Fig. (9). the amount of money contributed by a region is shown as a rounded whole number in the pie chart, while the actual value is listed in the data table (Table 1) that follows the chart.

Table 1. Data table listing specific earnings per region [By Mike Yi. From: https://chartio.com/learn/charts/pie-chart-complete-guide/] Google Images.

Region	Total Revenue
North	491 064.51
East	283 445.43
South	128 753.87
West	263 391.13

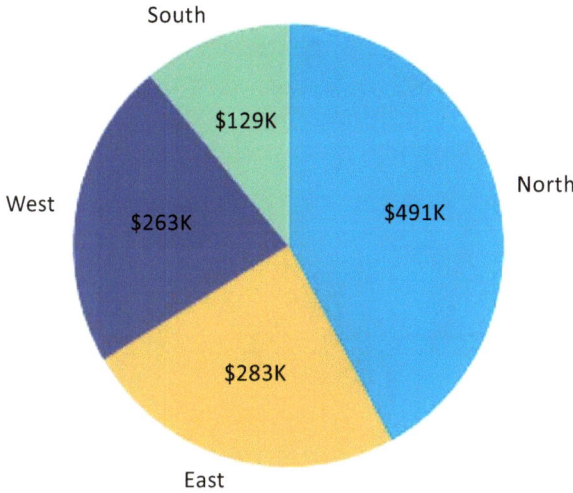

Fig. (9). Pie chart showing earnings by region [By Mike Yi. From: https://chartio.com/learn/charts/pie-char--complete-guide/] Google Images.

Pie charts are typically color-coded to differentiate the categorical variables shown in the slices, as well as roughly match the proportions of the depicted slices to the shown values. For example, a value of half of the total should be shown as roughly 50% of the pie. Pie charts also typically include annotations on the side of the chart that list the values represented by the slices, particularly if there are smaller slices of the pie where it would be difficult to show or list the given value [15]. An example of a pie chart with a sidebar is shown in Fig. (**10**).

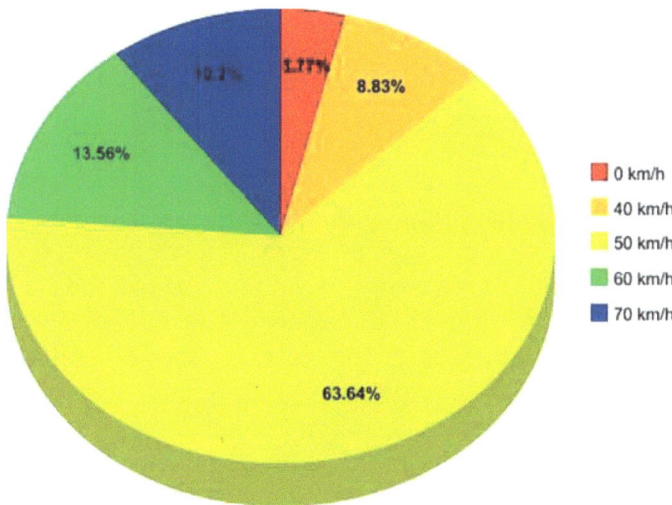

Fig. (10). Pie chart with sidebar [Source: Wikimedia Commons].

A special type of pie chart is the donut chart. A donut chart is essentially a pie chart with the center removed. Donut charts are used to represent portions of a whole as arcs rather than slices, which can sometimes better direct the reader to absorb the proportion of the contributions of each piece to the whole [12]. The donut chart can also show additional data, such as the total, using the empty space in the middle of the chart. An example of a donut chart is shown in Fig. (**11**).

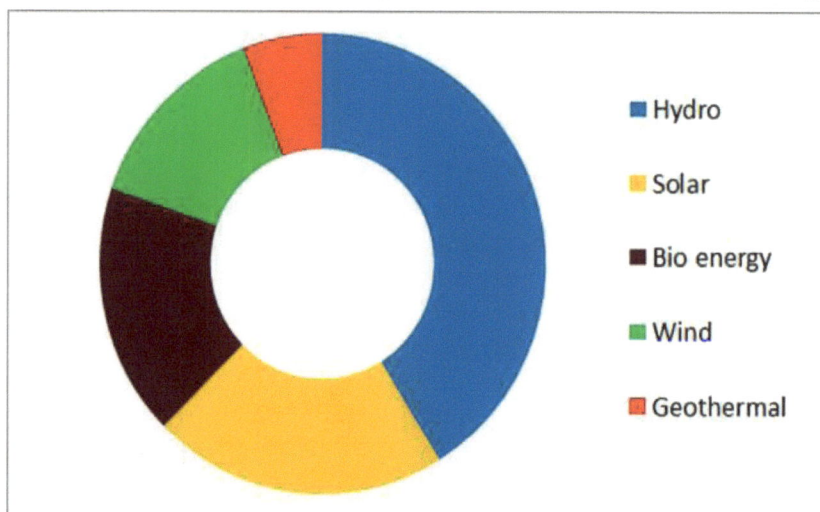

Fig. (11). Example donut chart [Source: Wikimedia Commons].

Another type of visual aid that is commonly used in statistics is the stem and leaf plot. Stem and leaf plots are commonly used for presenting qualitative data to visualize the shape of a distribution, like a histogram. A stem and leaf plot is constructed by sorting the data into ascending order [13]. The data is then divided into stems and leaves. Typically, the leaf is the last digit of the number, and the stem is the first digit of the number. If the values are large, the number may be rounded to a particular place, such as the hundredth place for the leaves, so the stem will then be the remaining digits to the left of the rounded place. For example, in a sorted set of values such as 44, 46, 47, 49, 63, 64, 66, 68, 68, 72, 72, 75, 76, 81, 84, 88, or 106, the first digit would be used as the stem, so 4, 6, 7, 8, 10 and the remaining digits would be used as the leaf. The stem and leaf plot is drawn as two vertical columns, with the stems on the left and the leaves on the right. No values are skipped, so even if there is no value for a stem and leaf, like for 50 in the example, the stem of 5 is still included. When a value is repeated, the leaf is also repeated. In the example, the value of 68 is repeated, so the leaf for 6 has 8 shown twice. For the given example, the stem and leaf plot is shown in Fig. (**12**).

```
Stem      |Leaf
         4|4 6 7 9
         5|
         6|3 4 6 8 8
         7|2 2 5 6
         8|1 4 8
         9|
        10|              6
```

Key: 6 |3 =63
Leaf unit: 1.0
Stem unit: 10.0

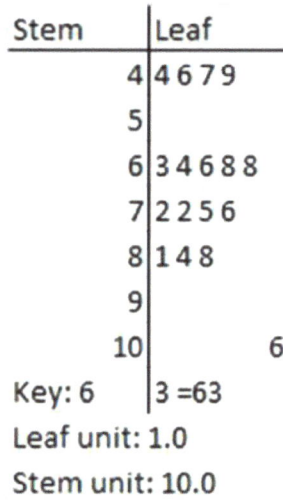

Fig. (12). Example stem and leaf plot (Created by author).

Stem and leaf plots help to visualize the trends of large data sets, such as showing if there is a bell curve or distribution associated with the data. Stem and leaf plots also show the density of the data and highlight outliers [13]. If there is a lot of data in a data set used for a stem and leaf plot, the data can be split into intervals to create an ordered stem and leaf plot. For example, in a data set such as reporting how often people drove to work over 10 days, the number of times each person drove was as follows (Table **2**):

5, 7, 9, 9, 3, 5, 1, 0, 0, 4, 3, 7, 2, 9, 8

Table 2. Number of drives to work in 10 days.

Stem	Leaf
0	0 0 1 2 3 3 4 5 5 7 7 8 9 9 9
Splitting the stems	

The leaves are crowded as each leaf is only based on a single reported digit. The organization of this stem and leaf plot does not give much information about the data. With only one stem, the leaves are overcrowded. If the leaves become too crowded, then it might be useful to split each stem into two or more components [14]. Thus, an interval 0–9 can be split into two intervals of 0–4 and 5–9. Similarly, a 0–9 stem could be split into five intervals: 0–1, 2–3, 4–5, 6–7 and 8–9.

The stem and leaf plot should then look like this Table **3**:

Table 3. Number of drives to work in 10 days.

Stem	Leaf
0(0)	0 0 1 2 3 3 4
0(5)	5 5 7 7 8 9 9 9

Note: The stem 0(0) means all the data within the interval 0–4. The stem 0(5) means all the data within the interval 5–9. By creating an ordered stem and leaf plot, the data becomes much clearer. Stem and leaf plots are most useful for spotting outlying values as well as obtaining the general shape of distribution within a dataset [16].

A hybrid type of plot that is frequently used in statistics is the dot plot. A dot plot is a way to show data that is a combination of a histogram and a stem and leaf plot. Dot plots are used to show individual data points within a set and how those points relate to each other on the scale of the given data set [15]. Dot plots are used for small data sets in which the data falls into discrete bins or categories. They are created by organizing the number of entries in a category into columns of dot, where each dot represents one count or vote in that category [22]. For example, a company tracking paint orders by the number of gallons could create a dot plot based on their customer orders for the week. In this example, there were five orders for one gallon of paint, three orders for two gallons, one order for four gallons of paint, two orders for five gallons, one order for six gallons, one order for seven gallons and one order for ten gallons of paint. This is shown using the dot plot in Fig. (**13**).

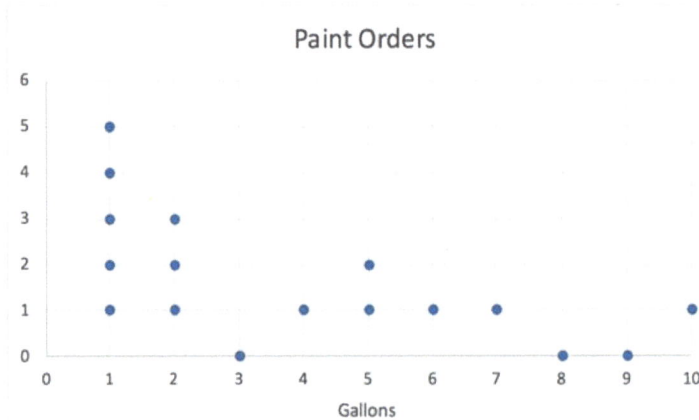

Fig. (13). Example dot plot of customer paint orders [Created by author].

Dot plots and frequency tables can be used to show the same data, just in different ways [15]. If a table is desired, the same data can be ordered to show the frequency of orders [22]. Using the dot plot in Fig. (**13**), it is easy to see that one gallon of paint was the most common order, followed by two gallons, five gallons and then four, six, seven and ten gallons of paint. Dot plots have been used to show data since 1884, and the Federal Reserve in the United States uses a dot plot each year to show the forecast for the central bank's key short-term interest rate [16]. The short-term interest rate is how much it will cost consumers to borrow money, and it is an important measure of the state of the American economy. Each dot in the Federal Reserve's dot plot represents the prediction of each official on the Federal Reserve [16]. The year is plotted on the x-axis, and the federal funds rate is plotted on the y-axis. A recent Federal Reserve dot plot is shown in Fig. (**14**).

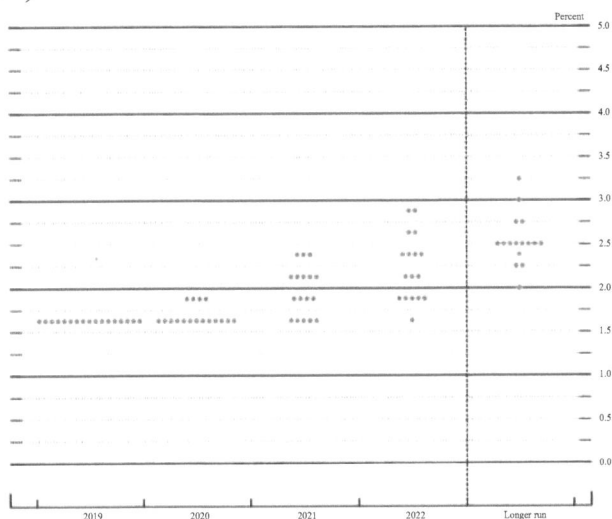

Fig. (14). Federal Reserve dot plot predictions for 2019- 2023. Public domain [16].

Another type of plot that is commonly used in statistics is the scatterplot. A scatterplot is used to show the relationship between two sets of data. Each dot represents an observation, while their positions on the x and y (horizontal and vertical axes, respectively) represent the values of the two variables [17]. Scatterplots are used to observe relationships between variables. They are best used to show or determine relationships between the two plot variables, or correlations [10, 18]. If the two variables show a positive relationship or correlation, a line can be plotted between them that increases vertically [17]. The better this line fits the given dots, the better the correlations of the two variables. If there is little correlation between the plotted variables, then the plot remains scattered, and any line plotted on the chart will pass through very few of the points. If the two variables show a negative correlation, then a line plotted between the points will decrease. If the two variables show a correlation that is strong, but do not show a linear relationship, the plot may take a curved or gaussian shape. The shape of each type of correlation within a scatterplot is shown in Fig. (**15**).

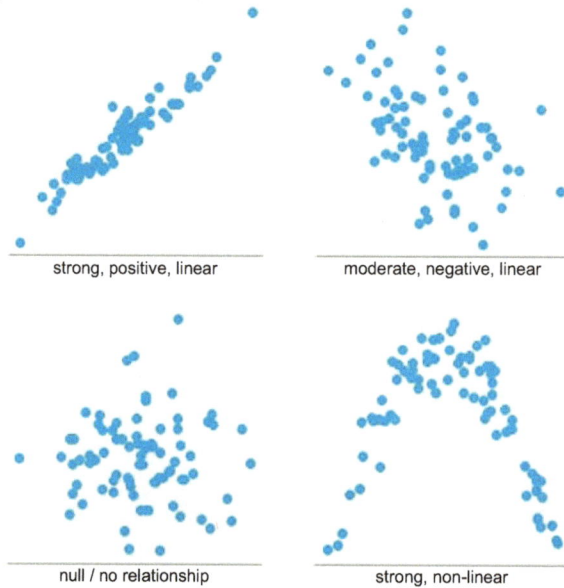

Fig. (15). Examples of different types of data correlations for scatterplots [By Mike Yi. From: https://chartio.com/learn/charts/what-is-a-scatter-plot/] Google Images.

Scatterplots can also be useful for identifying other patterns in a data set, for example, outliers or clusters of data. Gaps in values of data can also be seen when data is plotted using a scatterplot, as there will be a break in the clusters of data that are plotted. Fig. (**16**) shows other patterns in data that can be visible when using a scatterplot.

While scatterplots are very useful for plotting statistical relationships or showing other patterns in data, they are also very common issues that arise when they are used. Overplotting or showing too much data on a scatterplot can result when a very large dataset is used [19]. The data points shown may overlap, making it difficult to understand or see if there is truly a relationship between given data points, flooding the plot. This is easily solved by plotted a subset of the data or introducing dimensionality to the data [20]. Adding a layer of transparency can help clarify the relationships of the data points as a second dimension, or plotting the points as a two-dimensional histogram or heat map can also help clarify the relationships within the data [21]. Examples of each of these techniques appear in Fig. (**17**).

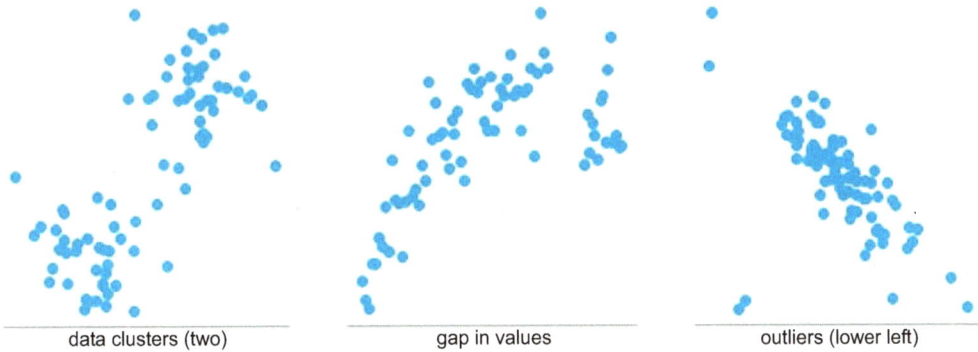

| data clusters (two) | gap in values | outliers (lower left) |

Fig. (16).   Other data patterns visible when using a scatterplot [By Mike Yi. From: https://chartio.com/learn/charts/what-is-a-scatter-plot/] Google Images.

Another potential issue that can occur when using scatterplots to visualize data is the interpretation of the data. When a relationship between data points arises from a scatterplot, it does not necessarily mean that changes in one variable are responsible for changes in another [12, 19]. Correlations in seemingly unrelated variables are referred to as spurious correlations. It is possible that the observed relationship is coincidental or driven by some third variable that affects both plotted variables. An example of a spurious correlation is shown in Fig. (**18**).

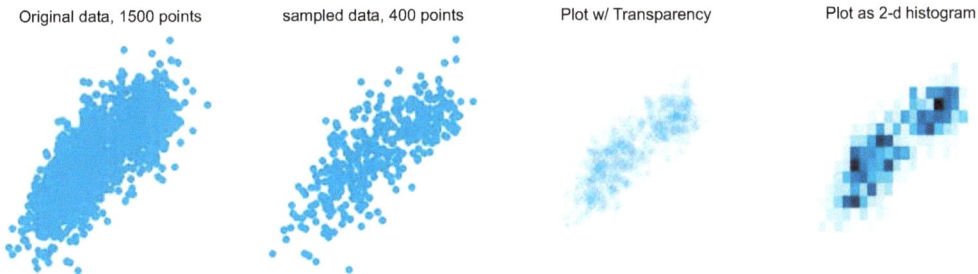

| Original data, 1500 points | sampled data, 400 points | Plot w/ Transparency | Plot as 2-d histogram |

Fig. (17).   Examples of clarifying data when it overplotting occurs [By Mike Yi. From: https://chartio.com/learn/charts/what-is-a-scatter-plot/] Google Images.

For graphing data that occurs over a set time interval, a time series graph is often used. This is a way for a data set to be visualized that shows progress and trends. The time interval range can be short or long, ranging from seconds to years [22]. A good example of a dataset that could be plotted on a time series graph would be the population of the United States over 50 years, or the approval rating of a president over the length of their term in office, such as for George W. Bush, as shown in Fig. (**19**). In a time series graph, the time is always shown on the horizontal axis. Data is then connected using a solid line, while missing data is plotted using a dashed line.

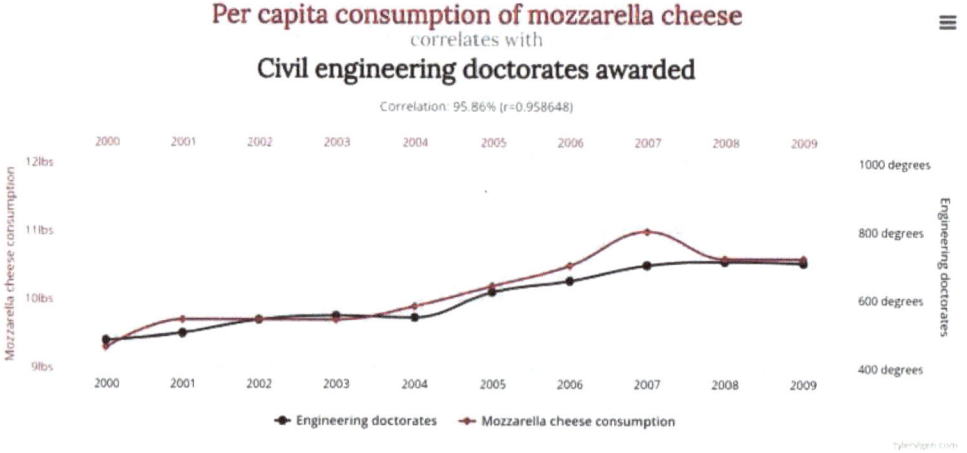

Fig. (18). Spurious correlation between civil engineering awarded doctorates and per capita consumption of mozzarella cheese. Courtesy tylervigen.com, used under Creative Commons license.

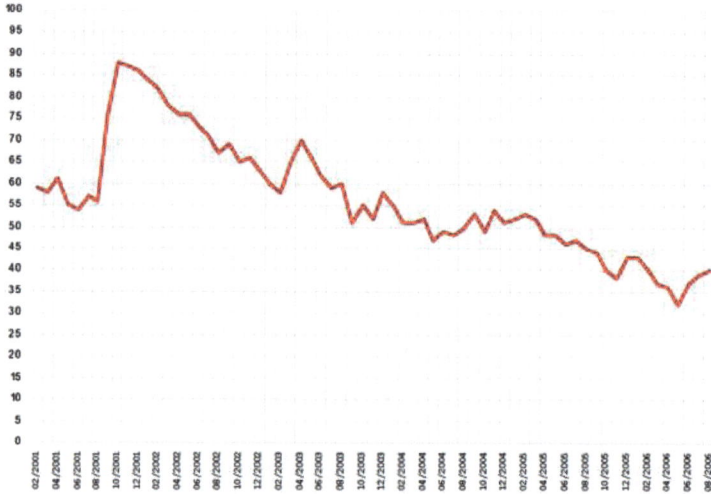

Fig. (19). Time series graph showing the approval rating of President George W. Bush over his time in office [Source: Wikimedia Commons].

Time series graphs can not only be used to show trends, but also to dig deeper into data sets by plotting multiple values over the same time interval [23]. For example, all the polling opinion options of a president over the same time interval can be plotted on the same graph to show the full breadth of the data [24]. This expanded time series graph is shown in Fig. (20).

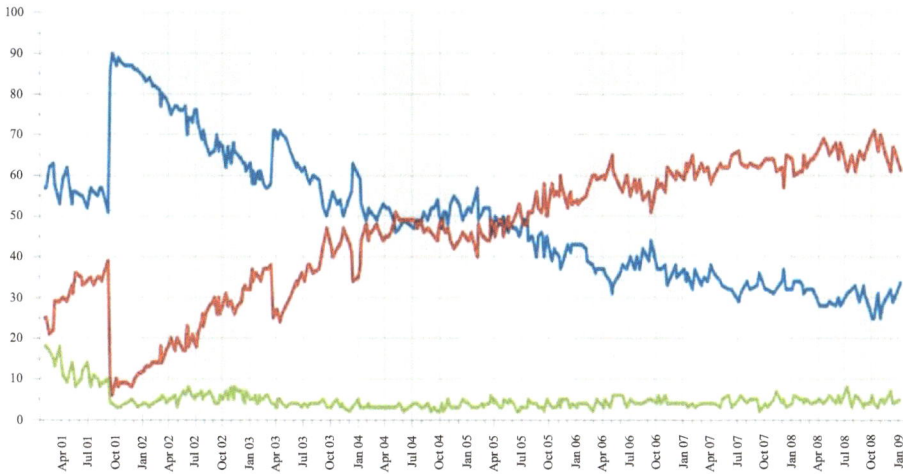

Fig. (20). George W. Bush opinion polling results.

Data visualization is a great tool for statistics as it allows individuals to analyze data, show trends and unearth critical information about a data set [25]. There are many different types of graphs that are utilized in statistics in addition to those listed here. Whether the data set is small or large, a graph or visualization technique can probably be applied. It is important to choose the best visualization for the data set as well as to utilize the data to construct the graph to its best advantage [25]. Sometimes, this means that only a subset of the data is visualized, or a certain time interval is graphed. This does not take away from the dataset, rather, it enhances the presentation of the information.

CONCLUSION

To conclude, it is important to acknowledge that the data visualization methods listed herein are only a few of the available options for showcasing data. Data visualization is a field in and of itself. There are many methods that can be used to show patterns, trends and other useful information concerning a particular dataset or population. The goal of data visualization is to improve understanding of statistical data *via* graphic or visual conveyance. Any method that effectively shows important information about a dataset that would otherwise be hidden or difficult to note can be an effective method of data visualization. It is also important to be cognizant that new data visualization tools are continually emerging within the field of statistics. New tools can provide a richer and deeper understanding of data as well as assist in analysis.

CHAPTER 3

Basic Concepts of Probability

Abstract: A commonly used statistical measure is the measurement of probability. Probability is governed by both additive and multiplicative rules. These rules determine if events are independent of one another or dependent on previous or related outcomes. Conditional probability governs events that are not independent of one another and helps researchers to better make predictions about future datasets.

Keywords: Conditional probability, Dependent, Independent, Probability.

INTRODUCTION

Probability is a commonly encountered statistical measure that helps to determine the likelihood of the outcome of a specific event. By understanding what the odds are of an event occurring, researchers can make further predictions about future datasets as well as to better understand collected data. The rules of probability govern the way that odds are generated as well as their interpretation. Conditional probability is the analysis of events that are not independent of one another and is frequently utilized to better understand collected datasets and make predictions about future outcomes.

A probability is a number that expresses the possibility or likelihood of an event occurring. Probabilities may be stated as proportions ranging from 0 to 1 and percentages ranging from 0% to 100%. A probability of 0 implies that an event cannot happen, while a probability of 1 suggests that an event is very likely to happen. A probability of 0.45 (45%) means that the event has a 45 percent chance of happening [26].

A study of obesity in children aged 5 to 10 seeking medical treatment at a specific pediatric clinic may be used to demonstrate the notion of likelihood. All children seen in the practice in the previous 12 months are included in the population (sample frame) described below [27].

Assume a polling company asks 1,200 people a series of questions to assess the percentage of all voters who support a certain bond issue. We would anticipate the percentage of supporters among the 1,200 polled to be similar to the percentage

Alandra Kahl

among all voters, but this does not have to be the case [28]. The survey result has a certain amount of unpredictability to it. We have faith in the survey result if the result is extremely likely to be close to the real percentage. If the percentage isn't likely to be near the population proportion, we shouldn't take the survey results too seriously. Our confidence in the survey result is determined by the possibility that the survey percentage is close to the population proportion [28]. As a result, we'd want to be able to calculate that probability. The problem of calculating it falls within the probability category, which we will look at in this chapter [28].

SAMPLES EVENTS AND THEIR PROBABILITIES

Sample Spaces

A common example of a random experiment is the rolling of a six-sided die; although all potential outcomes may be stated, the actual result on any specific trial of the experiment cannot be predicted with confidence [28]. When dealing with a situation like this, it is preferable to give a numerical number to each outcome (such as rolling a two) that indicates how often the event will occur. A probability would be assigned to any event or collection of outcomes, such as rolling an even number that shows the likelihood that the event will occur if the experiment is carried out similarly [28].

A random phenomenon's sample space is just the collection of all conceivable (basic) outcomes. Outcomes are the most fundamental things that can happen. When you roll a dice, for example, the potential outcomes are 1, 2, 3, 4, 5, and 6-- resulting in a sample space of {1,2,3,4,5,6} [28].

Following the specification of the sample space, a set of probabilities is given to it, either by repeated testing or through common sense. The outcomes are usually given the same probability; however, this is not always the case [28].

You commonly use letters to indicate outcomes, such as x or c, and then P to represent the probability of the occurrence (x). The following are the two rules that the probability assignments must follow: It must be true for any result x that

$0 < P(x) < 1$ (it is allowed for P(x) to equal 0 or 1 -- if P(x) = 0, it indicates that x virtually never occurs, and if P(x) = 1, it means that x practically always happens.) [28]

When all the probability of all the eventualities are added together, you get 1. (this means that your list of outcomes includes everything that can happen) [28].

Event

A random experiment is a method that generates a specific result that cannot be anticipated with confidence. A random experiment's sample space is the collection of all conceivable results. A subset of the sample space is an event [28].

If the result observed is an element of the set E, an event E is said to occur on a certain experiment trial [28].

Examples

Experiments are a part of life in almost every field of study. Special sorts of experiments are also addressed in probability and statistics. Consider the examples below.

Example 1

A coin is thrown. If the coin does not fall on the side, the experiment may have two alternative outcomes: heads or tails. It is impossible to indicate the result of this experiment at any time. You may throw the coin as many times as you like [29].

Example 2

A roulette wheel is a circular disc with 38 identical sections numbered 0 through 36, plus 00. The wheel is rolled in the opposite direction after a ball is rolled on its edge. Any of the 38 numbers, or a combination of them, maybe gambled on. You may also wager on a certain hue, red or black. If the ball falls in the number 32 sector, for example, everyone who bet on 32 or a combination of 32 wins, and so on. In this experiment, all potential results are known in advance, namely 00, 0, 1, 2,..., 36, yet the outcome is undetermined on every execution of the experiment, provided, of course, that the wheel is not fixed in any way. The wheel can be rolled an infinite number of times [29].

Example 3

A company makes 12-inch rulers. The experiment aims to measure the length of a ruler manufactured by the manufacturer as precisely as feasible. Because of manufacturing mistakes, it is impossible to tell the real length of the chosen ruler. However, the length will be between 11 and 13 inches, or between 6 and 18 inches if one wants to be careful [29].

Example 4

The lifespan of a light bulb made by a certain manufacturer is kept track of. In this scenario, the length of life of the light bulb chosen is unknown, but it will be somewhere between 0 and ∞ hours [29].

EXPERIMENT

The experiments discussed above have a few things in common. We know all potential results ahead of time; there are no surprises in store when the experiment is over. However, we do not know what the particular result will be on every given performance of the experiment; there is uncertainty about the outcome of each given performance. Furthermore, the experiment may be repeated under the same circumstances. A random (or statistical) experiment has these characteristics [29].

Definition:

A random (or statistical) experiment is one in which:

a. Any experiment performance yields an unknown outcome ahead of time [29].
b. The experiment may be repeated with the same results [29].

Example 5

Construct a sample space that defines all three-child households regarding the children's genders about their birth order [28].

Solution

Two possible results are "two boys then a girl," abbreviated as bbg, and "a girl then two boys," abbreviated as gbb. There are several consequences, and attempting to list them all may make it impossible to be certain that we have located them all unless we go in an organized manner. [28] "Tree Diagram for Three-Child Families," as illustrated in Fig. (**1**), provides a methodical technique.

The diagram was made in the following manner. Because the first kid may be either a boy or a girl, we draw two line segments from a beginning point, one terminating in a b for "boy" and the other in a g for "girl." Each of these two choices for the first kid has two possibilities for the second children, "girl" or "boy," so we draw two line segments from each of the g and b, one terminating in a b and the other in a g [28]. There are two options for the third kid for each of the four finishing places in the diagram now, so we repeat the procedure [28].

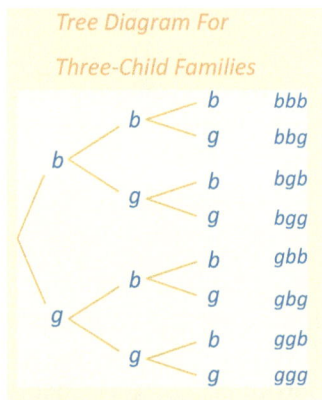

Fig. (1). Tree Diagram for Three-Child Families [28].

The tree's line segments are referred to as branches. Each branch's right-hand endpoint is referred to as a node. The final nodes are those on the far right; each corresponds to a result, as indicated in the diagram [28].

The eight results of the experiment can easily be read off the tree; therefore, the sample space is, reading from the top to the bottom of the tree's final nodes [28],

$$S = \{bbb, bbg, bgb, bgg, gbb, gbg, ggb, ggg\}$$

PROBABILITY

In a sample space S, the probability of a result e is a number p between 0 and 1, representing the chance that e will occur on a single trial of the related random experiment. The value p = 0 indicates that the result e is impossible, whereas p = 1 indicates that the outcome e is certain [28].

The probability of event A is equal to the sum of the probabilities of the various outcomes that make up that event. P is the abbreviation for it (A) [28].

The probability of event A is equal to the sum of the probabilities of the various outcomes that make up that event. It is denoted P(A) [28].

The following formula explains the substance of the probability of an event definition [28]:

If an event E is E = $\{e_1, e_2,..., e_k\}$, then

$$P(E) = P(e_1) + P(e_2) + ... + P(e_k)$$

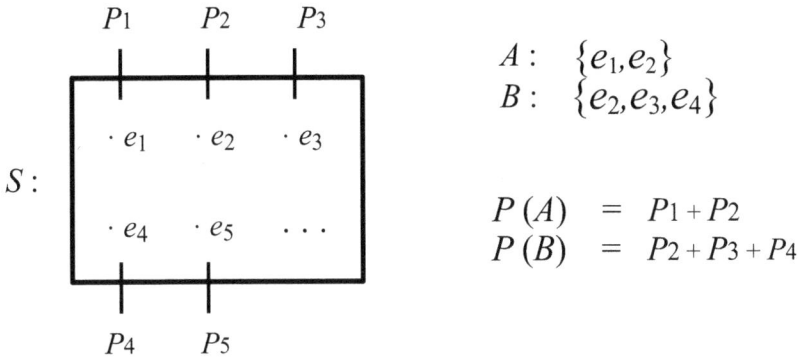

$$A: \quad \{e_1, e_2\}$$
$$B: \quad \{e_2, e_3, e_4\}$$

$$P(A) \quad = \quad P_1 + P_2$$
$$P(B) \quad = \quad P_2 + P_3 + P_4$$

Fig. (2). Whole sample space S is a guaranteed event.

Because the whole sample space S is a guaranteed event, the probability of all the outcomes must equal one (Fig. **2**) [28].

Probabilities are typically represented as percentages in everyday English. We may state, for example, that there is a 70% possibility of rain tomorrow or that the probability of rain is 0.70. We'll stick to this technique here, although, in the next computations, we'll use the form 0.70 instead of 70% [28].

Examples

Example 1

If each side of a coin has an equal chance of landing up, it is said to be "balanced" or "fair." For the experiment with a single fair coin flip, assign a probability to each result in the sample space [28].

Solution

The sample space is the set S={h,t},, with h for heads and t for tails being the outcomes. Each possibility is given a probability of 1/2 since all possibilities have to provide the same probabilities, which must sum up to 1 [28].

Example 2

If each side of the die has an equal chance of landing on top, it is said to be "balanced" or "fair." A For the experiment in which a single fair die is thrown, assign a probability to each occurrence in the sample space. Determine the occurrences' probability. T: "a number larger than two is rolled," and E: "an even number is rolled" [28].

Solution

The sample space is the set S={1,2,3,4,5,6} with outcomes labeled according to the number of dots on the top face of the die. Because six equally probable possibilities must sum up to one, each is given a probability of one-sixth of one percent [28].

Since E = {2,4,6}, *P(E)* = 1/6+1/6+1/6 = 3/6 = 1/2.

Since T = {3,4,5,6}, *P(T)*= 4/6 = 2/3.

Example 3

Two equally weighted coins are thrown. Determine the likelihood that the coins will match, *i.e.*, both land heads or land tails [28].

Solution

The probability theory does not teach us how to assign probabilities to events; it only tells us what to do with them after they have been allocated. Matching coins is the M=2h,2t, which has the probability P(2h)+P in sample space S. (2t). Matching coins is the occurrence M'=hh,tt in sample space S', with probability P(hh)+P. (tt) [28]. In the practical world, it shouldn't matter if the coins are similar or not; thus, we'd want to assign probabilities to the outcomes such that the values P(M) and P(M') are the same and best resemble what we find in actual physical trials with seemingly fair coins. Because we know that the events in S' are equally probable, we give a probability of 1/4 to each of them [28].

$$P(M') = P(hh) + P\ (tt) = 1/4 + 1/4 = 1/2$$

Similarly, from experience appropriate choices for the outcomes in S are:

$$P(2h) = 1/4\ P(2t) = 1/\ 4\ P(d) = \tfrac{1}{2}$$

which give the same final answer

$$P(M) = P(2h) + P(2t) = 1/4 + 1/4 = 1/2$$

When the sample space comprises a small number of equally likely possibilities, the previous three examples demonstrate how probabilities may be calculated simply by counting the number of options in the sample space that are equally probable. Consequently, individual outcomes of any sample space representing the experiment are inherently unequally likely in some situations; in these situations, probabilities cannot be computed simply by counting and must instead be computed using the computational formula specified in the definition of the probability of an event instead [28].

Example 4

A local high school's student body comprises 51 percent whites, 27 percent blacks, 11 percent Hispanics, 6 percent Asians, and 5 percent others. A student from this high school is chosen at random [28]. (Choosing "randomly" implies that each student has an equal chance of getting chosen). Determine the likelihoods of the following events:

B indicates that the student is black,

M indicates that the student is a minority (*i.e.*, not white), and

N indicates that the student is not black.

Solution

The experiment entails picking a kid randomly from the high school's student body. S={w,b,h,a,o} is an obvious sample space. P(w)=0.51, and similarly for the other outcomes, since 51 percent of the students are white and all students have the same probability of getting chosen. The following table summarizes this information [28]:

Outcome	w	b	h	a	o
Probability	0.51	0.27	0.11	0.06	0.05

a. Since B = {b}, P (B) = P (b) = 0.27.

b. Since M = {b, h,a, o},

$$P(M) = P(b) + P(h) + P(a) + P(0) = 0.27 + 0.11 + 0.06 + 0.05 = 0.49$$

c. Since N = {w, h, a, o},

$$P(N) = P(w) + P(h) + P(a) + P(0) = 0.51 + 0.11 + 0.06 + 0.05 = 0.73 \text{ [3]}$$

Example 5

The student population at the high school in Example 4 may be classified into the following 10 groups: 25% white male, 26% white female, 12% black male, 15%

black female, 6% Hispanic male, 5% Hispanic female, 3% Asian male, 3% Asian female, 1% male of other minorities combined, and 4% female of other minorities combined. A student from this high school is chosen at random. Determine the likelihoods of the following events [28]:

B denotes a black student.

MF denotes a minority female student, and FN denotes a female student who is not black.

Solution

S={bm,wm,am,hm,om,bf,wf,of,hf,af} is now the sample space. The following table, referred to as a two-way contingency table, summarizes the facts presented in the example:

Gender	Race / Ethnicity				
	White	**Black**	**Hispanic**	**Asian**	**Others**
Male	0.25	0.12	0.06	0.03	0.01
Female	0.26	0.15	0.05	0.03	0.04

a. Since *B= {bm, bf}, P (B) = P (bm) + P (bf)* = 0.12 + 0.15 = 0.27.

b. Since *MF = {bf, hf, af, of}*,

$$P(M) = P (bf) + P (hf) + P (af) + P (of) = 0.15 + 0.05 + -0.03 + 0.04 = 0.27$$

c. Since *FN = {wf, hf, af, of}*,

$$P(FN) = P (wf) + P (hf) + P (af) + P (of) = 0.26 + 0.05 + 0.03 + 0.04 = 0.38$$

COMPLEMENTS, INTERSECTIONS, AND UNIONS

Complement

The complement of an event A in a sample space S, abbreviated Ac, is the sum of all possible results in S that are not members of set A. It equates to denying any account of occurrence A that is expressed in words [28].

The complement of a set A is made up of anything in the original set A.

The complement is denoted by A', Ac, or A.

Examples

Example 1

E: "the number rolled is even" and T: "the number rolled is larger than two" are two occurrences associated with the experiment of rolling a single die. Find each person's complement [28].

Solution

The comparable sets of outcomes in the sample space S={1,2,3,4,5,6} are E={2,4,6} and T={3,4,5,6}. Ec={1,3,5} and Tc={1,2} are the complements [28].

"The number rolled is not even" and "the number rolled is not higher than two" are the complements in language. Of course, "the number rolled is odd" and "the number rolled is fewer than three" are simpler explanations [28].

What is the likelihood of good weather if there is a 60% risk of rain tomorrow? The obvious answer, 40%, is an example of the following general rule.

Probability Rule for Complements

$$P(A^c) = 1 - P(A)$$

When determining the likelihood of an occurrence directly is difficult, this formula comes in handy [28].

Example 2

Determine the likelihood that at least one head will emerge in five fair coin flips.

Solution

Lists of five hs and ts, such as tthtt and hhttt. Although it is laborious to name them all, counting them is not difficult. Consider using a tree diagram to do this. For the initial throw, you have two options. There are two options for the second throw for each of them, resulting in a total of 22=42×2=4 outcomes for two tosses. There are two options for the third throw for each of these four results, resulting in 4×2=8 outcomes for three tosses. Similarly, for four tosses, there are 8×2=16 outcomes, and for five tosses, there are 16×2=32 outcomes [28].

Let O stand for the occurrence "at least one heads." There are various methods to get at least one head, but only one way to get all tails. It is thus simple to write Oc={ttttt} because each of the 32 equally probable alternatives has a probability of 1/32, P(Oc)=1/32 and P(O)=1/32≈0.97 (about a 97 percent likelihood) [28].

Intersection of Events

The collection of all outcomes that are elements of both sets A and B is known as the intersection of events A and B, abbreviated as A ∩ B. It relates to using the word "and" to combine descriptions of the two occurrences [28].

When we remark that event A ∩ B happened, we refer to both A and B on one of the experiment's trials. Fig. **(3)** "The Intersection of Events" shows a graphic depiction of the intersection of events A and B in a sample space S. The darkened lens-shaped area that fits inside both ovals corresponds to the junction [28].

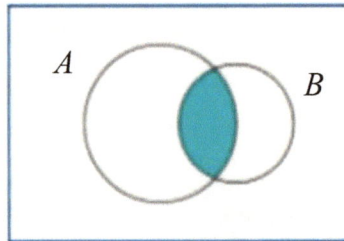

Intersection of Events

A and B

Fig. (3). The Intersection of Events [28].

Examples

Example 1

Locate the point of intersection. E ∩ T of the events E: "the number rolled is even" and T: "the number rolled is larger than two" in the experiment of rolling a single die [28].

Solution

S={1,2,3,4,5,6} is the sample space. Because the typical outcomes for E={2,4,6} and T={3,4,5,6} are 4 and 6, E∩T={4,6} [28].

"The number rolled is even and larger than two," explained the intersection in words. The only even and greater-than-two numbers between one and six are four and six, which match E ∩ T above [28].

Example 2

Only one die is rolled.

Assume the die is balanced. Calculate the chances that the rolled number is both even and bigger than two.

Assume the dice has been "loaded" such that P(1)=1/12, P(6)=3/12 and the four remaining possibilities are all equally probable. Calculate the chances that the rolled number is both even and bigger than two [28].

Solution

The sample space in both instances is S={1,2,3,4,5,6}, and the event in question is the intersection E∩T={4,6} from the preceding example.

a. Because all possibilities are equally probable because the die is fair, we can get P(E∩T)=2/6 by counting.
b. The information we have so far on the odds of the six scenarios is [28]

Outcome	1	2	3	4	5	6
Probability	1/12	p	p	p	p	3/12

Since $P(1) + P(6) = 4 / 12 = 1/3$ and the probabilities of all six outcomes add up to 1,

$$P(2)+ P(3) + P(4) + P(5) = 1-1/3=2/3$$

Thus $4p = 2/3$, so $p=1/6$. In particular $P(4) = 1/6$. Therefore

$$P(E∩T)=P(4)+P(6)= 1/6 +3/12 =5/12$$

Definition

If events A and B have no components in common, they are mutually exclusive.

The fact that A and B have no outcomes in common indicates that both A and B cannot occur on a single trial of the random experiment. As a result, the following rule emerges [28].

Probability Rule for Mutually Exclusive Events

Events A and B are mutually exclusive if and only if

$$P(A∩B)=0 \text{ [28]}$$

A and *Ac* are mutually exclusive in every case, while A and B might be mutually exclusive without complements.

Example

Find three options for event A in the experiment of rolling a single die such that events A and E: "the number rolled is even" are mutually incompatible.

Solution

Because E={2,4,6} and we don't want A to have many components in common with E, any event with an odd number will suffice. {1,3,5} (the complement Ec the odds), {1,3} and {5} are the three options [28].

Union of Events

The collection of all outcomes that are elements of one or both sets A and B labeled A υ B is known as the union of events A and B. It relates to using the word "or" to combine descriptions of the two occurrences [28].

The event A υ B happened signifies that either A or B happened on a certain experiment trial (or both did). Fig. (4) "The Union of Events" shows a graphic depiction of the union of events A and B in a sample space S. The darkened zone corresponds to the union [28].

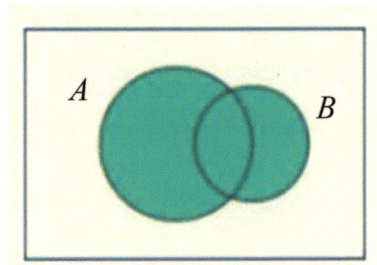

Fig. (4). The Union of Events [28].

Examples

Example 1

Find the union of the occurrences E: "the number rolled is even" and T: "the number rolled is higher than two" in the experiment of rolling a single die [28].

Solution

EυT={2,3,4,5,6} since the results in either E={2,4,6} or T={3,4,5,6} (or both) are 2, 3, 4, 5, and 6. It's worth noting that a result that appears in both sets, such as 4, is only reported once (although strictly speaking, it is not incorrect to list it twice) [28].

"The number rolled is even or more than two" is how the union is expressed in words. Every number between one and six, except one, is either even or bigger than two, corresponding to E ʊ T above [28].

Example 2

At random, a two-child family is chosen. Let B signify the case where at least one kid is a boy, D the case where the genders of the two children vary, and M the case where the genders of the two children are the same. B ʊ D and BʊM should be found [28].

Solution

S= {gg, bb, gb, bg} is a sample space for this experiment, with the first letter indicating the gender of the firstborn kid and the second letter indicating the gender of the second child. B, D, and M are the three events [28].

$$B = \{bb, bg, gb\} \ D = \{bg, gb\} \ M = \{bb, gg\} \ [28]$$

Because every event in D is already exists in B, the occurrences that are in at least one of the sets B and D are just the set B itself: BʊD= {bg, bb, gb}=B [28].

Every result in the whole sample space S belongs to at least one of the sets B and M; therefore,

$$B\upsilon M = \{bb, bg, gb, gg\} = S \ [28].$$

The following formula for estimating the probability of AʊB is the Additive Rule of Probability.

Additive Rule of Probability

$$P (A \cup B) = P(A) + P(B) - P(A \cap B) \ [28]$$

The following example demonstrates why the final term in the formula is required since we calculate the probability of a union both by counting and utilizing the formula [28].

Example 3

Two equally weighted dice are tossed. Determine the likelihoods of the following events:

- A four appears on both dice.
- a four appears on at least one dice

Solution

As with throwing two identical coins, real experience demands that we list outcomes as if we could differentiate the two dice for the sample space to contain equally probable outcomes. One of them may be red and the other green, for example. Then, as in the example below, every result may be labeled as a pair of numbers, with the first number being the number of dots on the top face of the green die and the second number being the number of dots on the top face of the red die [28].

$$
\begin{array}{cccccc}
11 & 12 & 13 & 14 & 15 & 16 \\
21 & 22 & 23 & 24 & 25 & 26 \\
31 & 32 & 33 & 34 & 35 & 36 \\
41 & 42 & 43 & 44 & 45 & 46 \\
51 & 52 & 53 & 54 & 55 & 56 \\
61 & 62 & 63 & 64 & 65 & 66
\end{array}
$$

a. There are 36 equally probable options, one of which corresponds to two fours, giving a pair of fours a 1/36 probability [28].

b. The table reveals that 11 pairings relate to the event in question: the six pairs in the fourth row (the green die shows a four) plus the extra five pairs in the fourth column (the red die shows a four), resulting in an answer of 11/36. Let AG symbolize the event that the green die is a four, and let AR denote the event that the red die is a four to show how the formula returns the same number. Then we can easily see that P(AG)=6/36 and P(AR)=6/36. Since AG∩AR={44}, P(AG∩AR)=1/36; this, of course, is the calculation in section (a). As a result, according to the Additive Rule of Probability [28],

$$P(A_G \cup A_R). = P(A_G) + P(A_R) - P(A_G\text{-}A_R) = 6/36 + 6/36 - 1/36 = 11/36$$

Example 4

Adults may use a tutoring service to prepare for high school equivalent examinations. Among all students needing assistance from the program, 63 percent require assistance in mathematics, 34 percent in English, and 27 percent want assistance in mathematics and English. What proportion of kids needs assistance in math or English [28]?

Solution

Consider picking a student at random, meaning that each student has an equal chance of getting chosen. Let M represent the event "the student need assistance in mathematics," and E represents the event "the student requires assistance in English." [3] P(M)=0.63, P(E)=0.34, and P(M∩E)=0.27 are the values presented. The Additive Rule of Probability is a formula for calculating the probability of a given event.

$$P(M \cup E) = P(M) + P(E) - P(M \cap B) = 0.63 + 0.34 - 0.27 = 0.70 \ [28]$$

Note how the naive logic that if 63 percent require math assistance and 34 percent need English aid, then 63 plus 34 or 97 percent need help in one or the other results in a much too huge. The proportion of students who need assistance in both disciplines must be reduced; otherwise, those who require assistance in both subjects would be counted twice, once for mathematics and again for English [28]. If the occurrences in the issue were mutually exclusive, the simple sum of probabilities would work since P(A∩B) is 0 and makes no difference.

Example 5

Volunteers for a disaster relief effort were categorized by their field of expertise (C: building, E: education, and M: medical) and their language skills (S: speaks a single language fluently, T: speaks two or more languages fluently). The following two-way categorization table displays the results:

Specialty	Language Ability	
	S	*T*
C	12	1
E	4	3
M	6	2

The first row of data indicates that 12 volunteers with a construction specialty communicate fluently in a single language, and 1 volunteer with construction specialty talks fluently in at least two languages. The same may be said for the other two rows.

A volunteer is picked at random, which means that everyone has the same chance of getting chosen. Calculate the likelihood that [28]:

His specialty is medicine, and he speaks two or more languages; his specialty is anything other than medicine, and he speaks two or more languages; his specialty is something other than medicine, and he speaks two or more languages [28].

Solution

When data is presented in a two-way categorization table, it is common to practice adding the row and column totals to the table, resulting in a new table that looks like this:

Specialty	Language Ability		Total
	S	T	
C	12	1	13
E	4	3	7
M	6	2	8
Total	22	6	28

a. The probability sought is $P(M \cap T)$. The table shows that there are 2 such people, out of 28 in all, hence $P(M \cap T) = 2 / 28 \approx 0.07$ or about a 7% chance.

b. The probability sought is $P(M \cup T)$. The third row total and the grand total in the sample give $P(M) = 8 / 28$. The second column total and the grand total give $P(T) = 6 / 28$. Thus, using the result from part (2),

$$P(M \cup T) = P(M) + P(T) - P(M \cap T) = 8/28 + 6/28 - 2/28 = 12/28 \approx 0.43$$

or about a 43% chance.

¢. This probability can be computed in two ways. Since the event of interest can be viewed as the event CUE and the events C and E are mutually exclusive, the answer is, using the first two row totals,

$$P(C \cup E) = P(C) + P(B) - P(C \cap B) = 13/28 + 7/28 - 0/28 = 20/28 \approx 0.71$$

On the other hand, the event of interest can be thought of as the complement M^c of M, hence using the value of $P(M)$ computed in part (b),

$$P(M^c) = 1 - P(M) = 1 - 8/28 = 20/28 \approx 0.71 \ [28]$$

CONDITIONAL PROBABILITY AND INDEPENDENT OCCURRENCES

Conditional Probability

The likelihood of one event happening in the context of one or more other occurrences is known as conditional probability [28].

Assume a fair die was rolled, and you're asked to estimate the likelihood that it was a five. Your answer is 1/6 because there are six equally probable possibilities. Assume, on the other hand, that you are given the extra information that the number rolled was an odd number before you decide. You'd alter your estimate of the chance of a five being rolled from 1/6 to 1/3 since there are only three odd numbers that may be rolled, one of which is five [28]. The conditional probability of A given B, abbreviated as P(A|B), is the updated probability that an event A has happened, considering the extra knowledge that another event B has certainly occurred on this trial of the experiment. The computational formula in the following definition may be derived from the logic used in this example [28].

The conditional probability of A given B, written P(A|B), is the probability that event A has happened in a random experiment trial for which event B has been determined. It may be calculated using the following formula:

Conditional Probability Rule:

$$P(A|B) = P(A \cap B)/P(B) \text{ [28]}$$

Examples

Example 1

The dice is rolled fairly.

a. Given that the number rolled is odd calculate the likelihood that it is a five.
b. Determine the likelihood that the rolled number is odd, assuming that it is a five.

Solution

The set S={1,2,3,4,5,6} is the sample space for this experiment, and it contains six equally probable events. Let F stand for "a five is rolled," and let O stand for "an odd number is rolled," such that.

$$F = \{5\} \text{ and } O = \{1,3,5\} \text{ [28]}$$

a. Because this is the first case, we already know the solution is 1/3. To use the formula in the definition to validate this, we must replace A (the event whose probability we want to estimate) with F and B (the event we know has happened) with O:

$$P(F|O) = P(F \cap O)/P(O)$$

Since $F \cap O = \{5\} \cap \{1,3,5\} = \{5\}$, $P(F \cap O) = 1/6$.

Since $O = \{1,3,5\}$, $P(O) = 3/6$.

Thus

$$P(F|O) = P(F \cap O)/P(O) = 1/6 \,/\, 3/6 = 1/3 \,[28]$$

b. This is an identical situation, but with F and O in inverted roles. Given that the number rolled was five, an odd number, the probability in question must be one. To apply the formula to this situation, replace A (the event whose probability we want to estimate) with O and B (the event we know has happened) with F:

$$P(O|F) = P(O \cap F)/P(F)$$

Obviously $P(F) = 1/6$. In part (a) we found that $P(F \cap O) = 1/6$. Thus

$$P(O|F) = P(O \cap F)/P(F) = 1/6 \,/\, 1/6 = 1 \,[28]$$

We don't need the computational formula in this case, and we don't need it in the following example when the information is provided in a two-way categorization table.

Example 2

Each participant was categorized according to gender and age at first marriage in a sample of 902 people under 40 who were or had previously been married. The findings are described in the two-way categorization table below, where the labels have the following meanings:

-	*E*	*W*	*H*	*Total*
M	43	293	114	450
F	82	299	71	452
Total	125	592	185	902

M: Stands for male.

F: Stands for female.

E: When she was first married, she was a teenager

W: When first married, in one's twenties

H: When first married, in one's thirties

The first row of statistics indicates that 43 men in the sample were first married in their teens, 293 men were first married in their twenties, 114 men were first married in their thirties, and 450 men were in the sample. The numerals in the second row are the same way. The final row of statistics indicates that 125 persons in the sample were married in their teens, 592 in their twenties, 185 in their thirties, and there were 902 people in the study. Assume that the proportions in the sample represent those in the population of all people under 40 who are married or have been married in the past. Assume that such a person is chosen at random [28].

a. Determine the likelihood that the person chosen was an adolescent at the time of their first marriage.
b. Given that the person is male calculate the likelihood that the person was an adolescent at their first marriage.

Solution

It's only natural to use E to express that the person chosen was an adolescent at the time of their first marriage, and M to represent the fact that the person chosen is male.

a. According to the table, 125/902 of the people in the sample were in their teens when they married for the first time. It represents the population's relative frequency of such persons, therefore $P(E)=125/902 \approx 0.139$, or almost 14% [28].

b. Because the individual chosen is a man, all females may be eliminated from consideration, leaving just the row in the table corresponding to men in the sample [3].

-	E	M	H	Total
M	43	293	114	450

43/450 of the men in the sample were in their teens when they married for the first time. The proportion of such persons in the male population, resulting in $P(E|M)=43/450 \approx 0.096$ or about 10% [28].

Example 3

Assume that the proportion of people who are both overweight and suffer from hypertension in an adult population is 0.09; The percentage of hypertensive persons who are not overweight but nonetheless have high blood pressure is around 0.11; The percentage of persons who are overweight but do not have hypertension are 0.02, and There is a significant fraction of persons who are neither overweight nor suffering from high blood pressure are 0.78. An adult is chosen at random from this group.

a. Determine the likelihood that the individual chosen has hypertension, given his weight.

b. Given that the chosen guy is not overweight calculate the likelihood of hypertension.

c. Compare the two probabilities you have discovered to determine if overweight persons are more likely to get hypertension.

Solution

Let H stand for the occurrence "the person chosen has hypertension." Let O stand for the occurrence "the individual chosen is overweight." The following contingency table may be used to arrange the probability information provided in the problem:

-	O	O^c
H	0.09	0.11
H^c	0.02	0.78

a. Using the formula in the definition of conditional probability,

$$P(H|O) = P(H \cap O) / P$$

b. Using the formula in the definition of conditional probability,

$$P(H|O^c) = P(H \cap O^c)/P(O^c) = 0.11/0.11 + 0.78 = 0.1236$$

c. $P(H|O) = 0.8182$ is over six times as large as $P(H|O^*) = 0.1236$, which indicates a much higher rate of hypertension among people who are overweight than among people who are not overweight. It might be interesting to note that a direct comparison of $P(H \cap O) = 0.09$ and $P(H \cap O^c) = 0.11$ does not answer the same question [28].

Independent Events

An independent event has no bearing on the likelihood of another event occurring (or not happening). In other words, the occurrence of one event has no bearing on the likelihood of another occurrence. In probability, independent occurrences are no different from independent events in real life. What color automobile you drive has nothing to do with where you work. Purchasing a lottery ticket has no bearing on the likelihood of having a kid with blue eyes [28].

Despite the fact that we often predict the conditional probability, P(A|B) to vary from the probability P(A) of A, this is not always the case (A). When P(A|B)=P(A), the probability of A is unaffected by the presence of B. The occurrence of event A is unrelated to the occurrence of event B [28].

It can be shown using algebra that the equality P(A|B)=P(A) holds if and only if the equality P(A∩B)=P(A).P(B) holds, which is true if and only if P(B|A)=P(B). The following definition is based on this.

Events A and B are independent if [28]:

$$P(A \cap B) = P(A) . P(B)$$

If A and B are not independent, then they are dependent.

The definition's formula has two practical but opposed applications [28]:

When all three probabilities P(A), P(B), and P(A∩B) can be computed, it is utilized to determine whether or not the occurrences A and B are independent:

- If $P(A \cap B) = P(A).P(B)$, then A and B are independent.
- If $P(A \cap B) \neq P(A).P(B)$, then A and B are not independent.

In a situation in which each of $P(A)$ and $P(B)$ can be computed, and it is known that A and B are independent, then we can compute P(A∩B) by multiplying together $P(A)$ and $P(B)$: $P(A \cap B) = P(A).P(B)$ [28].

Examples

Example 1

There is just one fair die rolled. Assume A={3} and B={1,3,5}. Are A and B independent?

Solution

We may calculate all three probabilities in this example: P(A)=16, P(B)=12, and P(A∩B)=P({3})=1/6. The occurrences A and B are not independent since the product P(A).P(B)=(1/6)(1/2)=1/12 is not the same as P(AB)=16 [28].

Example 2

In the preceding example, a two-way categorization of married or previously married persons under 40 by gender and age at first marriage yielded the table [28].

-	E	W	H	Total
M	43	293	114	450
F	82	299	71	452
Total	125	592	185	902

Determine the independence of the events F: "female" and E: "was a teenager at first marriage."

Solution

According to the table, 452 of the 902 adults in the sample were female, 125 were teens in their first marriage, and 82 were female teenagers in their first marriage, indicating that

$$P(F) = 452/902 \quad P(E) = 125/902 \quad P(F \cap E) = 82/902$$

Since,

$$P(F) \cdot P(E) = 452/902 \cdot 125/902 = 0.069$$

Is not the same as

$$P(F \cap E) = 82/902 = 0.091$$

We conclude that the two events are not independent [28].

Example 3

Many illness diagnostic tests do not test for the disease itself but rather for a chemical or biological result of the disease and are not completely trustworthy. The sensitivity of a test refers to the likelihood that it will be positive when given to someone who has the condition. The bigger the detection rate and the lower the false-negative rate, the better the sensitivity [28].

Assume that the sensitivity of a diagnostic technique for determining if a person has a certain ailment is 92%. Two separate labs use this approach to test for the illness in a person who has it.

a. What is the likelihood of both test results being positive?
b. What is the likelihood of at least one of the two test results being positive?

Solution

a. Let A_1 denote the event "the test by the first laboratory is positive" and let A_2 denote the event "the test by the second laboratory is positive." Since A_1 and A_2 are independent,

$$P(A_1 \cap A_2) = P(A_1) \cdot P(A_2) = 0.92 \times 0.92 = 0.8464$$

a. Using the Additive Rule for Probability and the probability just computed,

$$P(A_1 \cup A_2) = P(A_1) + P(A_2) - P(A_1 \cap A_2) = 0.92 + 0.92 - 0.8464 = 0.9936$$

Example 4

The likelihood that a diagnostic test for a disease will be negative when delivered to someone who does not have the ailment is specific. The smaller the false positive rate, the better the specificity.

Assume that a diagnostic procedure's specificity for determining if a person has a given ailment is 89 percent.

a. This method is used to test a person who does not have the condition. What is the likelihood that the test will provide a positive result?
b. Using this technique, a person who does not have the condition gets tested by two independent laboratories. What is the likelihood of both test results being positive [28]?

Solution

a. Assign B to the occurrence "the test result is positive." The complement of B is that the test result is negative, with a probability

$$P(B) = 1 - P(B^c) = 1 - 0.89 = 0.11 \text{ [28]}$$

b. Let B1 represent the event "the first laboratory's test is positive," and B2 represent the event "the second laboratory's test is positive." By part (a) of the example, B1 and B2 are independent.

$$P(B_1 \cap B_2) = P(B_1) \cdot P(B_2) = 0.11 \times 0.11 = 0.0121 \text{ [28]}.$$

The notion of independence may be used in a variety of situations. For example, three events A, B, and C are independent if $P(A \cap B \cap C) = P(A)P(B)P(C)$. Keep in mind that, as with merely two events, this is not a formula that is always correct, but it is correct when the events in question are independent [28].

Probability in Tree Diagrams

For computing combined probabilities for sequences of occurrences, probability trees are helpful. It allows you to visually map out the odds of many different scenarios without utilizing sophisticated probability formulae.

The following example shows how to use a tree diagram to represent probability and solve a problem.

Example

A jar holds ten marbles, seven of which are black and three white. Two marbles are pulled without being replaced, meaning the first one is not returned before the second is drawn [28].

Is there a chance that both marbles are black?

What's the chance that one of the marbles is black?

What's the chance that at least one of the marbles is black?

Solution

Fig. (**5**) "Tree Diagram for Drawing Two Marbles" depicts a tree diagram for the condition of drawing one marble after the other without replacement. The circle and rectangle will be discussed later, so don't worry about them right now.

On the first draw, you may choose a white marble (3 out of 10) or a black marble (3 out of 10). Given that the event corresponding to the node on the left end of the branch has happened, the number on each remaining branch represents the chance of the event corresponding to the node on the right end of the branch occurring. P(B2|B1), This is the top branch, which is connected to the two Bs, and it represents the event "the first pebble retrieved is black," and B2 represents the event "the second marble generated is black," respectively. Because 9 marbles remain after pulling a black marble, 6 of which are black, this chance is 6/9 [28].

Tree Diagram for Drawing Two Marbles

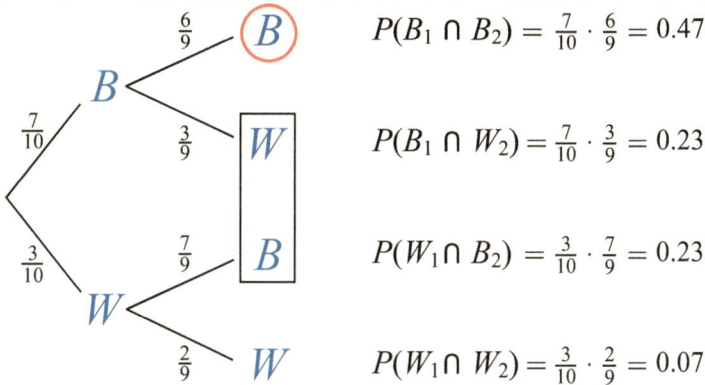

$$P(B_1 \cap B_2) = \tfrac{7}{10} \cdot \tfrac{6}{9} = 0.47$$

$$P(B_1 \cap W_2) = \tfrac{7}{10} \cdot \tfrac{3}{9} = 0.23$$

$$P(W_1 \cap B_2) = \tfrac{3}{10} \cdot \tfrac{7}{9} = 0.23$$

$$P(W_1 \cap W_2) = \tfrac{3}{10} \cdot \tfrac{2}{9} = 0.07$$

Fig. (5). Tree Diagram for Drawing Two Marbles [28].

The number to the right of each final node is calculated as indicated, based on the idea that if the Conditional Rule for Probability's formula is multiplied by P(B), the result is [3]

$$P(B \cap A) = P(B) \cdot P(A|B) \text{ [28]}$$

a. The event "both marbles are black" is B1∩B2, and it corresponds to the circled top-right node in the tree. As a result, it is 0.47, as mentioned there.
b. The two nodes of the tree highlighted by the rectangle coincide to the occurrence "exactly one marble is black," which is represented by the text "essentially one marble is black." These two nodes correspond to occurrences that are mutually incompatible with one another: black preceded by white is incompatible with white followed by black. Due to this, using the Additive Rule for Probability, we essentially combine the two probabilities next to those vary locations since deleting them from the accumulation would result in the value of 0. Thus, getting precisely one black marble in two trials has a chance of 0.23+0.23=0.46 [28].

c. It is the event "at minimum one pebble is black" that corresponds to the three nodes of the tree that are contained inside either the circle or the rectangle. To account for the fact that the events that equate to these nodes are simultaneously exclusive, we add the probabilities next to them, just as we did in section (b). Thus, getting at least one black marble in two attempts has a chance of 0.47+0.23+0.23=0.93.

Of fact, using the Probability Law for Complements, this result might have been calculated more quickly by subtracting the probability of the complimentary event, "two white marbles are chosen," from 1 to get 1−0.07=0.93.

As this example demonstrates, calculating the probability for each branch is rather simple since we know everything that has occurred so far in the chain of steps. The following are two broad ideas that arise from this example:

Principles

1. The product of the numbers on the unique route of branches that goes to that node from the start determines the probability of the event corresponding to that node on a tree [28].

2. If numerous final nodes relate to an event, the probability is calculated by summing the numbers adjacent to those nodes [28].

CONCLUSION

Probability is one of the most commonly used statistical measures and there are many methods and rules which govern its usage, only a few of which are treated here. With this statistical element, it is often useful to describe it in terms of set notation as we have done. More broadly, probability can be measured in terms of outcomes, such as heads or tails of a coin toss or the likelihood that a particular event will occur. Probability measurements and set notations assist researchers in analyzing future outcomes based on existing dataset as well as to understand patterns within data. Venn diagrams and union diagrams are the most frequently used tools to visualize probability.

CHAPTER 4

Discrete Random Variables

Abstract: Discrete random variables are variables that can only take on a countable number of discrete or distinct variables. Only a finite number of values can be obtained for the result to be a discrete random variable. Discrete variables can be visualized or understood as a binomial distribution. The outcome of a binomial distribution is how often a particular event occurs in a fixed number of times or trials.

Keywords: Binomial distribution, Outcome, Random.

INTRODUCTION

Random Variables

Unknown variables and functions that assign values to each of an experiment's results are referred to as random variables and functions, respectively. For the most part, random variables are denoted by letters, and they may be divided into two categories: discrete variables, which are variables with specified values, and continuous variables, which are variables that can take on either value within a certain range of values. Statistical connections between random variables are often seen in econometric and regression analysis, where they are utilized to discover statistical links between variables [30].

Understanding Random Variables

Random variables are exploited in probability and statistics to measure the results of a random event, and as a result, they might have a wide range of possible values. Random variables must be measured to be used, and real numbers commonly represent them. After three dice are rolled, the letter X may be selected to symbolize the number that comes up as a consequence. In this scenario, X may be three (1+1+1), eighteen (6+6+6), or anything in between, since the greatest number on a die is six, the lowest number is one, and the highest number on a die is six [30].

A random variable differs from an algebraic variable because it is not predictable. A parameter in an algebraic equation is an unknown value but may be determined

Alandra Kahl

by calculation. With the equation 10 + x = 13, we can see that we can compute the precise number for x, which happens to be 3. On the other contrary, a random variable has a set of possible values, and any of those values might result in the desired outcome, as shown by the dice in the preceding example [30].

Various features, such as the average price of an asset over a specific period, the return on investment after several years, the projected turnover rate at a firm during the next six months, and so on, may be assigned random variables in the corporate world. Whenever a risk analyst wants to evaluate the likelihood of an undesirable event happening, they must include random variables in their risk models. These factors are provided *via* scenario and sensitivity analysis tables, which risk managers utilize to make judgments on risk mitigation strategies [30].

Types of Random Variables

A random variable can be a discrete variable or a continuous variable. Discrete random variables have a finite number of different values that may be calculated. Consider the following experiment: a coin is thrown three times, and the results are recorded. If X represents the number of times the coin has shown up heads, then X is a discrete random variable that can have the quantities 0, 1, 2, and 3 when the coin comes up heads (from no heads in three successive coins tosses to all heads). There is no other potential value for X [30].

A continuous random variable may represent any value inside a particular range or period, and the variable can take on an endless number of potential values. An experiment in which the quantity of rainfall in a city is measured over a year, or the height of a random group of 25 individuals, would be an example of a continuous random variable [30].

As an illustration of the latter, consider the case in which Y represents a random variable representing the average height of a random group of 25 persons. You will discover that the final output is a continuous number since height maybe 5 feet or 5.01 feet or even 5.0001 feet in height. To be sure, there is an endless number of different heights to choose from [30]!

Random variables have probability distributions, which describe the possibility that any possible values will be seen for that variable. Assume that the random variable, Z, is the number that appears on the top face after it has been rolled once in a row. As a result, the potential values for Z will be 1, 2, 3, 4, 5, and 6, respectively. Each of these numbers has a 1/6 chance of being the value of Z since they are all reasonably plausible to be the value of Z [30].

In the case of a dice throw, the chance of obtaining a 3 or P (Z=3) is 1/6, as is the likelihood of receiving a 4 (Z=2), or any other number, on all six sides of the die. It is important to note that the total of all probability equals one [30].

Example of Random Variable

A coin flip is an excellent demonstration of a random variable in statistical analysis. For example, suppose you had a probability distribution in which the outcomes of a random event are not equally likely to occur. If the random variable, Y, represents the number of heads we get from tossing two coins, then Y might be one of three values: 0, one, or two. We may get no heads, one head, or both heads, depending on how the dice fall on a two-coin toss [30].

On the other extreme, the two coins land in four possible configurations: TT, HT, TH, and HH. As a result, the probability of receiving no heads is 1/4 since we only have one opportunity of getting no heads (*i.e.,* two tails [TT] when the coins are tossed). In a similar vein, the chance of receiving two heads (HH) is one in every four. It is important to note that receiving one head can occur twice: in HT and TH. As a result, P (Y=1) = 2/4 = 1/2 in this example [30].

Discrete Random Variables

When it comes to discrete variables, they are variables that can "only" be represented by specified integers on the number line [30].

Discrete variables are often denoted by capital letters: X, Y, Z, *etc.*

X is a random variable, and the probabilities in the probability distribution of X must meet the following two scenarios [30]:

- Each probability P(x) must be between 0 and 1 to be valid: $0 \leq P(x) \leq 1$.
- In this case, the total of all probability is one: P(x) =1.

Example:

A variable that can only contain integers or a variable that can only contain positive whole numbers are both examples of this kind of variable [30].

In the case of discrete variables, they may either take on an unlimited number of values or be restricted to a fixed number of possible values [30].

For example, the number we get when rolling a die is a discrete variable, and it can only have one of these values: 1, 2, 3, 4, 5, or 6 [30].

In Sweden on November 25th, for example, the number of raindrops that fall across a square kilometer might have an "infinite" number of values, a discrete variable with an "infinite" number of values.

Even though this amount is not mathematically infinite, it is common to practice and permissible to presume that it is.

Examples of Probability Distributions for Discrete Random Variables (DRV)

Example 1

A fair coin is thrown twice to determine its worth. Let X be the total number of heads that have been seen [31].

a. Create a probability distribution for the variable X.
b. Calculate the likelihood that at least one head will be spotted.

Solution

The available values for X are 0, 1, and 2, respectively. Each of these values corresponds to an event in the sample space S=hh, ht, th, tt of equally probable outcomes for this experiment: X = 0 to tt, X = 1 to ht, th, and X = 2 to hh. The sample space S=hh, ht, th, tt is comprised of the following events: The chance of each of these occurrences occurring, and hence the associated value of X, may be calculated simply by counting, as shown in the following table [31].

x	0	1	2
$P(x)$	0.25	0.50	0.25

The above table is a probability distribution of X.

a. X = 1 and X = 2 are mutually exclusive occurrences, and the event X 1 is the union of these two events. "At least one head" is the event X 1, which is the union of these two mutually exclusive events. Thus [31]

$$P(X \geq 1) = P(1) + P(2) = 0.50 + 0.25 = 0.75$$

Fig. (1) is a histogram that graphically demonstrates the probability distribution in terms of a probability distribution. Fig. (1) "Probability Distribution for Tossing a Coin Twice" is an example of the probability distribution.

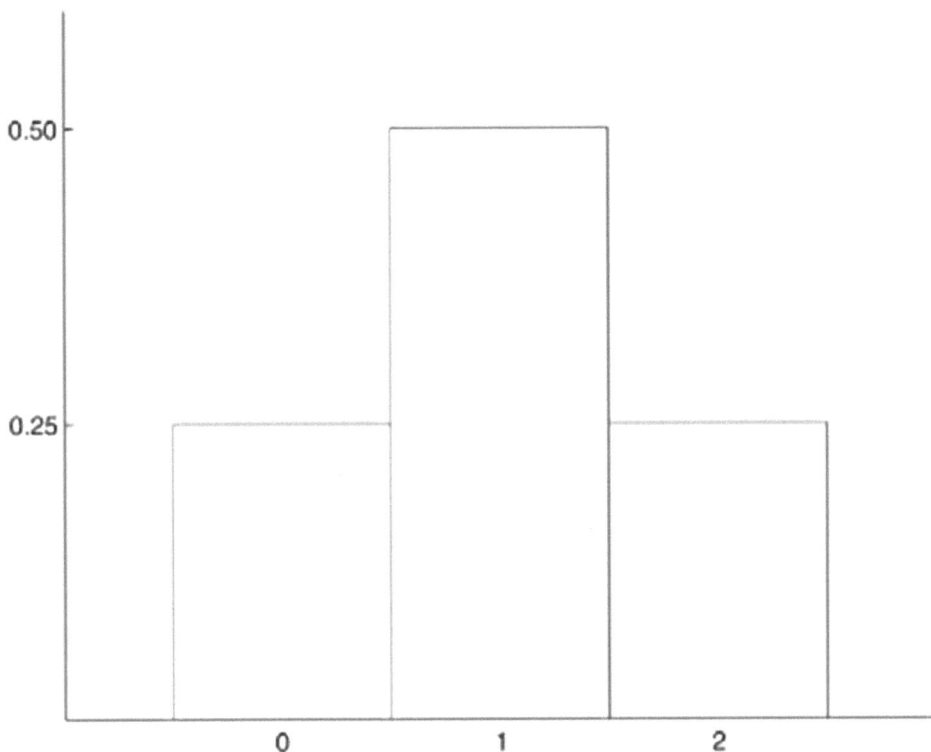

Fig. (1). Probability Distribution for Tossing a Coin Twice [31].

Example 2

A pair of fair dice is rolled to determine the outcome. The letter X denotes the total of the number of dots on the upper faces [31].

a. Establish a probability distribution for the variable X.
b. Calculate P (X 9).
c. Calculate the likelihood that X will have an even value.

Solution

The specimen space of equally probable outcomes is denoted by the symbol [31]

$$
\begin{array}{cccccc}
11 & 12 & 13 & 14 & 15 & 16 \\
21 & 22 & 23 & 24 & 25 & 26 \\
31 & 32 & 33 & 34 & 35 & 36 \\
41 & 42 & 43 & 44 & 45 & 46 \\
51 & 52 & 53 & 54 & 55 & 56 \\
61 & 62 & 63 & 64 & 65 & 66
\end{array}
$$

a. The digits 2 through 12 represent the range of potential values for X. P (2) =1/36 because X = 2 corresponds to the occurrence 11. As X = 3 corresponds to the occurrence 12, 21, P (3) =2/36. Continuing in this manner, we arrive at the table [31].

x	2	3	4	5	6	7	8	9	10	11	12
$P(x)$	$\frac{1}{36}$	$\frac{2}{36}$	$\frac{3}{36}$	$\frac{4}{36}$	$\frac{5}{36}$	$\frac{6}{36}$	$\frac{5}{36}$	$\frac{4}{36}$	$\frac{3}{36}$	$\frac{2}{36}$	$\frac{1}{36}$

The above table is probability distribution of X.

b. There are four mutually exclusive events in this set, and they are X = 9, X = 10, X = 11, and X = 12. The event X = 9 is the union of these four mutually exclusive occurrences. Thus [31]

$$
P(X \geq 9) = P(9) + P(10) + P(11) + P(12) = \frac{4}{36} + \frac{3}{36} + \frac{2}{36} + \frac{1}{36} = \frac{10}{36} = 0.2\overline{7}
$$

c. Please remember that X may take six distinct even values but only five possible odd values. This is essential to keep in mind before jumping to the conclusion that X's probability takes an even value must be 0.5. We do the math [31].

$$
\begin{aligned}
P(X \text{ is even}) &= P(2) + P(4) + P(6) + P(8) + P(10) + P(12) \\
&= \frac{1}{36} + \frac{3}{36} + \frac{5}{36} + \frac{5}{36} + \frac{3}{36} + \frac{1}{36} = \frac{18}{36} = 0.5
\end{aligned}
$$

Fig. (2). "Probability Distribution for Tossing Two Dice" is a histogram that graphically demonstrates the probability distribution for tossing two dice.

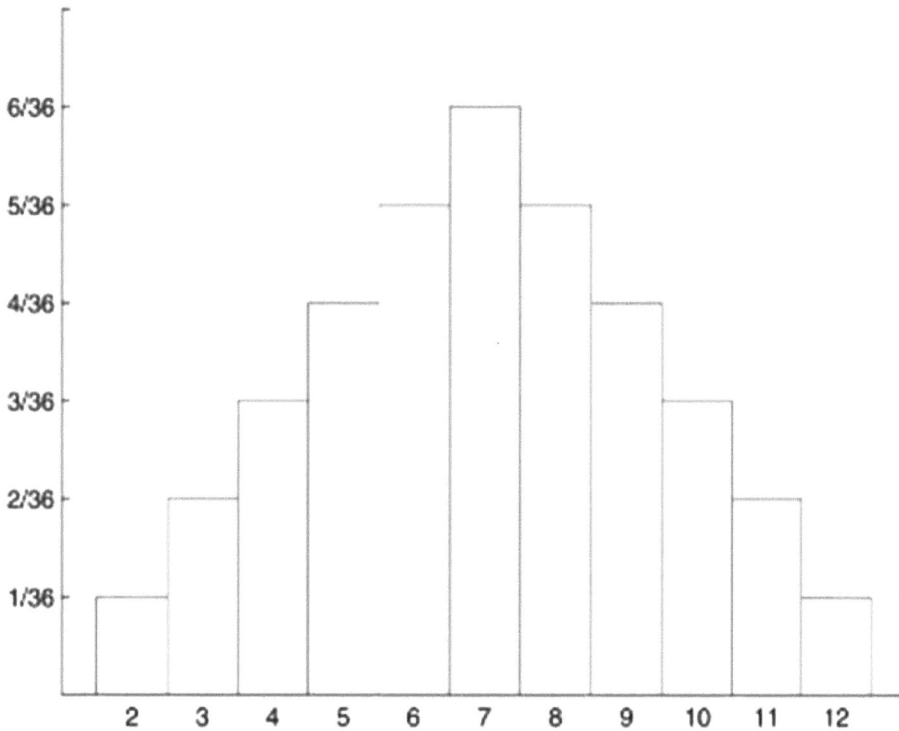

Fig. (2). Probability Distribution for Tossing Two Dice [31].

The Mean and Standard Deviation (SD) of Discrete Random Variables

The mean (also known as the following expression) of a discrete random variable X is the number of times the variable has occurred [31].

$$\mu = E(X) = \Sigma x\, P(x)$$

It is possible to interpret the mean of a random variable as the average of the values assumed by the random variable throughout a series of repeated trials of the experiment.

Examples

Example # 1

Calculate the mean of the discrete random variable X, whose probability distribution is as follows [31]:

x	−2	1	2	3.5
$P(x)$	0.21	0.34	0.24	0.21

Solution

The formula in the description yields the following [31]:

$$\mu = \Sigma x\, P(x)$$
$$= (-2) \cdot 0.21 + (1) \cdot 0.34 + (2) \cdot 0.24 + (3.5) \cdot 0.21 = 1.135$$

Example # 2

Every month, a service group in a big city holds a raffle to raise funds for its causes. One thousand raffle tickets are being offered at the cost of one dollar each. Each player has an equal probability of winning the game. The first prize is worth $300, the second prize is worth $200, and the third award is worth $100. Let X represent the net benefit resulting from purchasing a single ticket [31].

a. Construct a probability distribution for the variable X.
b. Calculate the likelihood of winning any money from purchasing a single lottery ticket.
c. Determine the predicted value of X and evaluate the significance of that value.

Solution

a) A ticket that wins the first prize will result in a net gain to the purchaser of $300 minus the $1 spent to acquire the ticket, resulting in X = 300 1 = 299 as the net gain to the purchaser. P (299) = 0.001 since there is only one such ticket. The following probability distribution is obtained by using the same "income minus outgo" method for the second and third prizes as well as for the 997 lost tickets [31]:

x	299	199	99	−1
$P(x)$	0.001	0.001	0.001	0.997

b) In the case that a ticket is picked to win one of the rewards, let W represent the event. Making use of the table [31]

$$P(W) = P(299) + P(199) + P(99) = 0.001 + 0.001 + 0.001 = 0.003$$

c) In the definition of anticipated value, the formula is used to calculate [31]

$$E(X) = 299 \cdot 0.001 + 199 \cdot 0.001 + 99 \cdot 0.001 + (-1) \cdot 0.997 = -0.4$$

The negative figure indicates that, on average, one loses money in the market. Specifically, if someone were to purchase tickets regularly, he would lose 40 cents for every ticket bought on average, even though he might sometimes win.

Example # 3

If a person falls into a certain risk category, an insurance firm will offer them a term life insurance policy for $200,000 that will last one year for a premium of $195. Figure out how much a single policy will be worth to the organization if a person in this risk category has a 99.97 percent probability of living for one year [31].

Solution

Let X represent the net profit realized by the corporation due to the sale of one such insurance. Alternatively, the insured individual may survive the whole year or die before it is through. According to the "income minus outgo" concept, the value of X in the former situation is 195 0; in the latter case, the value of X is 195 200,000=199,805. As the likelihood in the first scenario is 0.9997 and the probability in the second situation is 10.9997=0.0003, the probability distribution for X is as follows [31]:

x	195	$-199{,}805$
$P(x)$	0.9997	0.0003

Therefore,

$$E(X) = \Sigma x\, P(x) = 195 \cdot 0.9997 + (-199{,}805) \cdot 0.0003 = 135$$

Now and then (in fact, three times out of 10,000), the firm suffers a significant financial loss on a policy, but on average, it makes $195, which, according to our E(X) calculation, equates to a net gain of $135 on every insurance sold on average.

Variance of Discrete Random Variables

The variance, σ2, of a discrete random variable X is the number of occurrences of the variable [31].

$$\sigma^2 = \Sigma(x - \mu)^2 P(x)$$

This, according to algebra, is the same as the formula

$$\sigma^2 = \left[\Sigma x^2 P(x)\right] - \mu^2$$

As a result, the standard deviation of a discrete random variable X is equal to the square root of its variance, and may be calculated using the following formulas [31]:

$$\sigma = \sqrt{\Sigma(x - \mu)^2 P(x)} = \sqrt{\left[\Sigma x^2 P(x)\right] - \mu^2}$$

As measurements of the variability in the values assumed by the discrete random variable X during repeated trials in a statistical experiment, X's variance and standard deviation may be understood as follows: The units on the standard deviation correspond to the units on X [31].

Characteristics and Notations

It is common to see the distribution of a discrete random variable represented in a table, but a graph or a mathematical formula may also represent it. The following are the two most important features it should have [32]:

Each probability is a number between zero and one, excluding zero and one. The total probability is one hundred percent [32].

The probability mass function of a DRV provides information on the likelihood of a single value occurring. This is denoted by the notation P (X = x). This is sometimes called a probability distribution function, which is incorrectly termed (PDF) [32].

The cumulative distribution function of a DRV shows you how likely a value will be less than or equal to a certain amount. This is represented as P (X ≤ x) in terms of notation [32].

A probability distribution function represents a pattern. To execute the required computations, you must first attempt to fit a probability issue into a pattern or distribution that you have created. These distributions are tools that may be used to make addressing probability questions more straightforward. Each distribution has its own set of traits that distinguish it from others. Understanding the properties of the various distributions allows you to differentiate between them [32].

Binominal Distribution

A value can take one of two independent values under a particular set of inputs or assumptions, and the binomial distribution expresses this likelihood [33].

For the binomial distribution to work, it is necessary to make the following assumptions:

o There is only one consequence result for each trial.

o Each trial has the same chance of success.

o Each trial is mutually unique or independent of the previous trials.

Understanding Binominal Distribution

Compared to continuous distributions, such as the normal distribution, the binomial distribution is a typical discrete distribution in statistics utilized in many applications. This is because the binomial distribution only checks two possible outcomes, commonly expressed as 1 (for success) or 0 (for a failure), depending on the number of trials in the data. When given a success probability p for each trial, the binomial distribution may be used to predict how many times you will succeed in x trials in n trials [33].

When each trial has the same chance of achieving a given value, the binomial distribution describes the number of trials or observations. With the binomial distribution, you may calculate the likelihood of witnessing a specific number of successful outcomes in a particular number of trials [33].

Often used in social science statistics as a cornerstone for models of dichotomous outcome variables, such as whether a Republican or Democrat will win a forthcoming election or whether a participant will die within a specified period of time, the binomial distribution is also known as the binomial distribution of a binary outcome variable [33].

Analyzing Binominal Distribution

It is possible to analyze the expected value of a binomial distribution, or its mean, by multiplying the number of trials (n) by the chance of success (p), which is written as (n p) [33].

For example, the anticipated value of the number of heads in a hundred trials of head and stories is 50, which is equal to (100 * 0.5) times 100. One other typical use of the binomial distribution is evaluating the probability of achievement for a throw shooter in basketball, where 1 indicates a successful attempt and 0 indicates a failed attempt [33].

To compute the binomial distribution formula, use the following formula [33]:

$$P_{(x:n,p)} = {}_nC_x \times p^x(1-p)^{n-x}$$

where:

The number of trials is represented by the letter n. (occurrences)

The number of successful trials is represented by the letter X.

The chance of success in a single trial is denoted by the letter p.

When considering the binomial distribution, the mean is np (1 p), and when considering the variance, it's also np (1 p). Whenever p is equal to 0.5, the distribution is symmetric around the population's mean. A left-skewed distribution is seen when the probability density function (p) exceeds 0.5. When p is less than 0.5, the distribution is skewed to the right and *vice versa* [33].

The binomial distribution is the sum of a sequence of several independent and identically distributed Bernoulli trials, each of which has a probability of being true. If an experiment is conducted according to the Bernoulli principle, the results are considered randomized and can only have two potential outcomes: success or failure [33].

The flipping of a coin is an example of a Bernoulli trial because the outcome of each trial can only be one of two possibilities (heads or tails), each success has the same probability (the chance of flipping a head is 0.5), and the results of one trial do not impact the outcomes of another. The Bernoulli distribution is a specific instance of the binomial distribution in which the number of trials n = 1 and the number of observations n = 1 [33].

Criteria for Binominal Distribution

In the case of specified requirements being satisfied, the chance of an event occurring is represented by a binomial distribution. If you want to utilize the binomial probability formula, you need to follow the principles of binomial distribution. These rules must be followed throughout the procedure [34].

Trials With a Fixed Number of Trials

The procedure under inquiry must have a predetermined number of trials that cannot be changed throughout the course of the research or the analysis. During the analysis, each experiment must be carried out consistently, despite the fact that each trial may provide a different result [34].

In the binomial probability formulation, the letter "n" represents the number of trials that have taken place. Coin flips, free throws, wheel spins, and other similar activities are examples of fixed trials. The number of times that each experiment will be done is known from the beginning of the process. In the case of a coin flipping ten times, each flip of the coin represents a trial [34].

Randomized Controlled Trials

Besides being independent of one another, the trials of a binomial probability must also be independent of one another. To put it another way, the result of one experiment should not have any bearing on the outcome of future trials.

Because of the risk of having trials that are not totally independent of one another when using certain sampling techniques, the binomial distribution should only be utilized when the population size is big compared to the sample size [34].

Tossing a coin or rolling dice are two examples of independent trials that may be performed. It is vital to note that the initial event is completely independent of all future occurrences when flipping a coin [34].

There is a Fixed Chance Of Success

A binomial distribution requires that the likelihood of success stay constant throughout all of the trials under consideration. For example, in the case of a coin toss, the chance of flipping a coin is either 12 or 0.5 for every trial we do since there are only two potential outcomes [34].

In certain sampling approaches, such as sampling without alternatives, the chance of success from one trial to the next may vary from one trial to the next, depending on the methodology used. Consider the following scenario: 1,000 pupils in a class with 50 guys. Picking a guy from that demographic has a 0.05 percent chance of happening [34].

There will be 49 guys among the 999 pupils participating in the upcoming trial. According to the data, picking a male in the following experiment has a chance of 0.049 percent. It demonstrates that the probability from one trial to the next will differ somewhat from the probability from the preceding trial in successive trials [34].

Two Possible Outcomes Are Mutually Exclusive

There are only two possible exclusive outcomes in binomial probability: success or failure. However, although success is often considered a positive phrase, it may also be used to signify that the result of the experiment coincides with your definition of success, regardless of whether the outcome is good or negative [34].

Consider this scenario: A company gets a shipment of lamps with a high number of breakages. The company may define success for the trial as every light that does not include a piece of shattered glass. A failure may be described as a situation where there are no shattered glasses in the lights [34].

For the sake of this example, examples of broken lights may be used to demonstrate success as a means of proving that a significant fraction of the lamps in the consignment is broken and that there is a limited likelihood of receiving a shipment of lights with zero breakages [34].

Examples of Binominal Distributions

For the binomial distribution, multiply the probability of success raised to the power of how many successes there were by the likelihood of failure raised to the power of how many failures there were by the difference between how many successes there were and how many tries there were. After that, raise the product by the sum of the number of tries and the number of successes, if any [33].

Consider the following: imagine that a casino has developed a new game in which players may put bets on the number of heads or tails that will appear in a set number of coin flips. Consider the following scenario: a player wishes to put a $10 wager on the outcome of 20 coin flips: there will be precisely six heads. The participant is interested in calculating the likelihood of this happening, and as a result, they use the binomial distribution computation [33].

To compute the likelihood, divide the number by six and multiply by 0.50. The result is: (20! / (6! * (20 - 6)!)) * (0.50) *(6) * (1 - 0.50) (20 - 6). As a result, the likelihood of precisely six heads happening in a coin flip is 0.037, or 3.7 percent, when 20 coins are flipped. The anticipated value was ten heads; hence, the player placed a terrible bet.

Example

According to the most recent police statistics, 80 percent of all small crimes remain unsolved, and at least three of these petty crimes are perpetrated in your town each month. The three offenses are unrelated to one another and are committed on separate occasions. In light of the information provided, what is the likelihood that one of the three crimes will be solved [34]?

Solution

Finding the binomial probability requires first determining if the circumstance fulfills each of the four rules of the binomial distribution, which are as follows:

o The number of fixed trials (n) is three (Number of petty crimes)

o There are two mutually incompatible possibilities in this scenario (solved and unsolved)

o The likelihood of success (p) is 0.2. (20 percent of cases are solved)

o Yes, there have been independent trials.

Next:

We calculate the likelihood that one of the crimes will be solved in each of the three separate investigations. It is represented as follows:

Trial 1

Consists of three trials: the first, the second, and the third.

Solved 1st, unsolved 2nd, and 3rd

= 0.22 x 0.08 x 0.8

= 0.128 0.128

Trial 2

Unsolved first, solved second, and unsolved third

= 0.8 × 0.2 x 0.8

= 0.128

Trial 3

Unsolved first, unsolved second, and solved third

= 0.8 x 0.8 x 0.2

= 0.128

The total (for the three trials) is as follows:

= 0.128 + 0.128 + 0.128

the value of 0.384

Alternatively, we might make use of the information included in the binomial probability formula, as shown below [35].

$$P = \binom{N}{x} p^x (1-p)^{N-x}$$

Where [35]

$$\frac{n}{x} = \frac{n!}{x!(n-x)!}$$

In the above equation, x = 1 and n = 3. The equation gives a probability of 0.384.

Cumulative Binominal Probability

A cumulative binomial probability is a likelihood that a binomial random variable will fall inside a particular range, as defined by the probability distribution (*e.g.,* greater than or equal to a stated lower limit and less than or equal to a stated upper limit) [6].

Regarding coin tosses, we could be interested in the cumulative binomial probability of getting 45 or fewer heads in 100 tosses, as shown in the following example. In this case, the total of all individual binomial probabilities would be the result [35].

$$b(x \le 45; 100, 0.5) =$$
$$b(x = 0; 100, 0.5) + b(x = 1; 100, 0.5) + \ldots + b(x = 44; 100, 0.5) + b(x = 45; 100, 0.5)$$

Negative Binominal Distribution

In statistics, a negative binomial experiment is a statistical research that exhibits the following characteristics:

The experiment comprises a total of x trials that are repeated.

Each trial has two potential results, and each has two possible outcomes. One of these results is a success, while the other is referred to as a failure.

P denotes that the chance of success is the same on every try, regardless of the circumstances [35].

The events are independent of one another, which means that the result of one experiment does not affect the results of the other trials.

The experiment is carried out until a total of r successes are detected, where r is a predetermined number [35].

Take, for example, the statistical experiment described below. You flip a coin repeatedly, keeping track of how many times the coin falls on your heads each time. You continue to flip the coin till it has landed on heads five times in a row. This is a negative binomial experiment due to the following reasons:

The experiment is made up of a series of trials that are repeated. Flip a coin until it has landed on heads five times.

Each trial has two potential outcomes: either heads or tails, depending on the number of trials conducted.

At 0.5 every trial, the likelihood of success remains constant throughout the experiment.

There are no dependencies between the trials, which means that obtaining heads on one trial does not affect whether or not we get heads on subsequent trials.

The experiment continues until a certain number of successes have been achieved; five heads were achieved in this example.

Notations

When discussing negative binomial probability, it is useful to use the following notation.

x: The number of trials necessary to get r successes in a negative binomial experiment is denoted by the x.

r: The proportion of successes in the negative binomial experiment is represented by the letter r.

P: The likelihood of achieving success on an individual try.

Q: What is the likelihood of failure on a specific trial?

A: (This is the same as 1 - P.)

B*(x; r, P): Negative binomial probability is the chance that an x number of trials will negative binomial experiment will result in the rth success on the xth trial when the probability of success on a particular trial is equal to the probability of success on the previous trial. P.

nCr: nCr is the number of combinations of n items that can be made by taking r things at a time.

n!: The factorial of the number n (also known as n factorial).

Consider the following scenario: a negative binomial study comprises of x trials and yields r successes. For example, if the chance of success on a single trial is P, then the negative binomial probability is as follows [35]:

$$b^*(x, r, P) = {}_{x-1}C_{r-1} * P^r * (1 - P)^{x-r}$$

$$b^*(x, r, P) = \{ (x-1)! / [(r-1)!(x-r)!] \} * P^r * (1 - P)^{x-r}$$

The Mean of Negative Binominal Distribution

If we characterize the mean of the negative binomial distribution as the average number of trials necessary to obtain r successes, then the mean is equal to [35]:

$$\mu = r / P$$

Where μ is the mean number of events, r is the number of achievements, and P is the probability of success on any particular trial.

Example

Bob is a high school basketball player who likes to shoot the ball. He is a 70% free-throw shooter. This indicates that his chance of making a free throw is 0.70 out of 100. What is the likelihood that Bob will make his third free throw on his fifth attempt throughout the course of the season?

Solution

This is an example of a negative binomial trial. The probability of success (P) is 0.70, the number of trials (x) is 5, and the number of successes (r) is 3. The probability of success (P) is 0.70, the number of trials (x) is 5, and the number of successes (r) is 3.

In order to answer this issue, we must input these numbers into the negative binomial equation [35].

$$b^*(x; r, P) = {}_{x-1}C_{r-1} * P^r * Q^{x-r}$$
$$b^*(5; 3, 0.7) = {}_4C_2 * 0.7^3 * 0.3^2$$
$$b^*(5; 3, 0.7) = 6 * 0.343 * 0.09 = 0.18522$$

CONCLUSION

In conclusion, a random variable is a numerical description of the outcome of a statistical experiment. There are many types of random variables as well as ways to calculate and visualize their relevancy to statistics. Random variables can be discreted or obtained by counting, and there are variables obtained by measuring, which are continuous random variables. Discrete random variables result in binomial distributions, which are simple ways to understand the probability of a success or failure experiment that is repeated multiple times.

CHAPTER 5

Continuous Random Variables

Abstract: Distributions of data give important information about the dataset, both to researchers and to analysts. Continuous random variables are those variables whose outcomes are measured instead of random variables that are counted. These variables can be handled using probability density functions and cumulative distribution functions. Both have appropriate times and usage cases.

Keywords: Binomial, Continuous, Normal.

INTRODUCTION

The distribution of data within a dataset is important information as adequately understanding the shape of a dataset assists researchers to both draw better conclusions and visualize their results during analysis. A binomial distribution is used when researchers want to understand the consequences of two independent outcomes, whereas a Poisson distribution is used to better understand the results of independent trials with the additional inclusion of time within the dataset. Distributions are powerful analyses of data that researchers can use to add depth and understanding to the analysis of their datasets.

A random variable X is seen as being continuous if and only if the probability that its recognition will fall within the interval [a,b] can be represented as an integral [36]:

$$P(X \in [a,b]) = \int_{a}^{b} f_X(x)dx$$

Whereas integrated part

$$f_X : \mathbb{R} \to [0,\infty)$$

Is known as probability density of function X.

It is yet critical to note that an integral is used to compute the area under a curve. If you want to define a continuous variable, you may consider its integral as the surface under the probability density function in the period between a and b (Fig. **1**) [36].

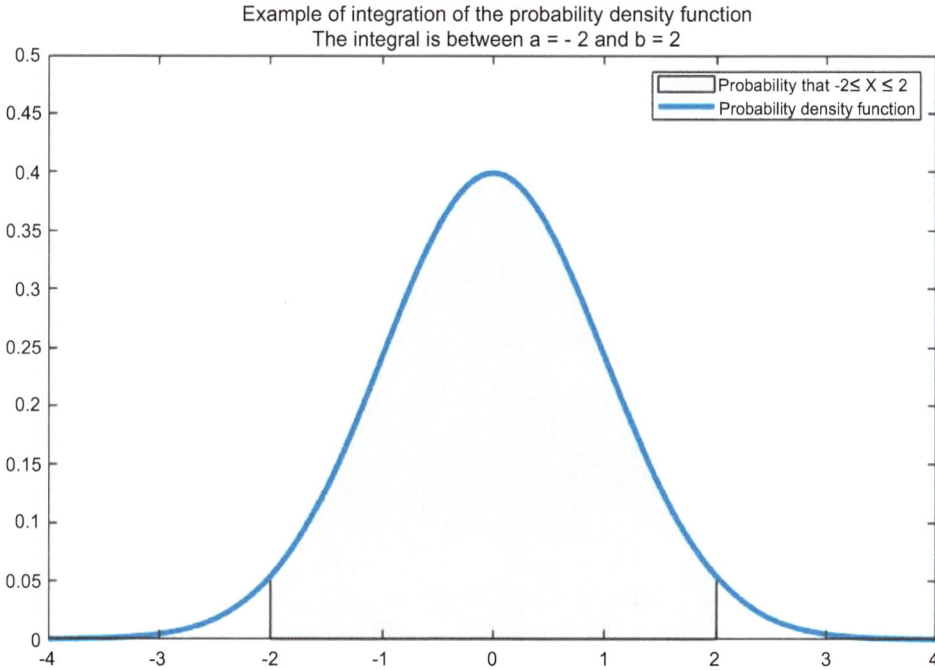

Fig. (1). Example of integration of the probability density function [36].

An illustration of a continuous random variable is one that has two primary characteristics [36]:

• The set of possible values it may take is not countable, and the cumulative distribution function of the variable is determined by integrating a probability density function [36].

Intervals are given probabilities based on the data.

The first thing to observe about the definition preceding is that the assignment of probabilities describes the distribution of a continuous variable to intervals of integers rather than the assignment of frequencies. Consider the fact that the distribution of a discrete integer is defined by assigning probabilities to single numbers, as opposed to the former [36, 37].

The uniform distribution (Fig. **2**) is a continuous probability distribution that deals with occurrences with an equal chance of occurring. When solving issues with a uniform distribution, keep in mind whether the data includes or excludes endpoints.

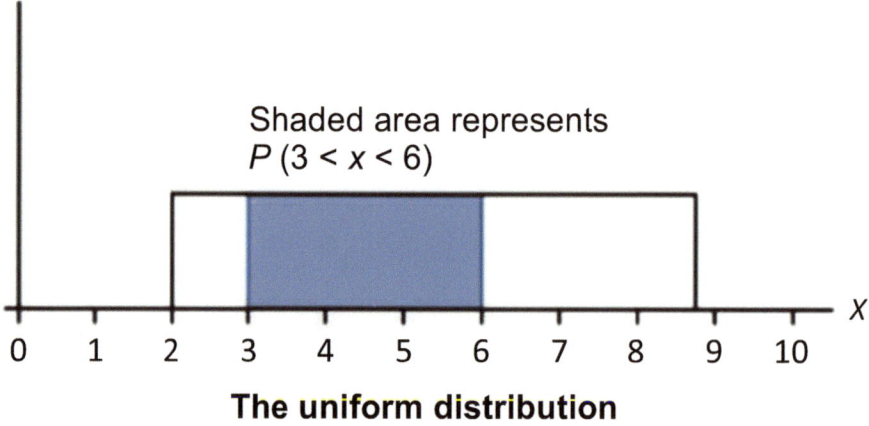

The uniform distribution

Fig. (2). The uniform distribution [37].

The exponential distribution (Fig. **3**) is often used to calculate the length of time before a certain event happens. The length of time (starting now) before an earthquake happens, for example, has an exponential distribution. Other examples are the duration of long-distance business phone conversations in minutes, and the time a vehicle battery lasts in months. It may also be shown that the value of the change in your pocket or handbag follows a roughly exponential distribution.

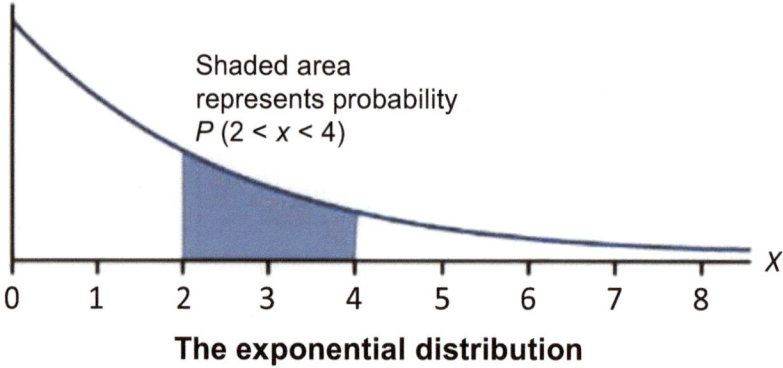

The exponential distribution

Fig. (3). The exponential distribution [37].

Exponential distributions are widely employed in product dependability estimates or determining how long a product will survive.

Probability Distribution of Continuous Random Variable

In contrast to the probability distribution of a discrete random variable, the probability distribution of a continuous random variable cannot be determined in the same manner as that of a discrete random variable. The likelihood of every single result occurring for a discrete random variable may be computed accurately using a probability mass function for a discrete random variable. For continuous random variables, the likelihood of any single event happening is zero, and the probability of any single outcome not occurring is one. To assess the likelihood of any result occurring during a certain timeframe, it is necessary to calculate the probability of each outcome occurring. This may be accomplished *via* a probability density function (PDF) [38].

Properties

A curve is a graph of a continuous probability distribution. The area under the curve represents probability.

The probability density function is the name of the curve (abbreviated as PDF). The curve is represented by the symbol $f(x)$. $f(x)$ is the graph's corresponding function; we use the density function $f(x)$ to create the probability distribution graph.

A separate function, termed the cumulative distribution function, determines the area under the curve (abbreviated as CDF). When calculating probability as an area, the cumulative distribution function is utilized.

• The results are measured rather than counted.
• The total portion under the curve and above the x-axis is one.
• Individual x values are not used to calculate probability; instead, intervals of x values are used.
• The chance that the random variable X is in the interval between the values c and d is given by $P(c < x < d)$. $P(c < x < d)$ is the area under the curve to the right of c and the left of d, above the x-axis.
• $P(x = c) = 0$ There is 0% chance that x will take on anyone's value. There is no width and hence no area (area = 0) below the curve, above the x-axis, and between x = c and x = c. The probability is also 0 since the probability is equal to the area.
• As probability is equal to the area, $P(c < x < d)$ is the same as $P(c \leq x \leq d)$.

We'll use geometry, mathematics, technology, or probability tables to discover the region that reflects probability. For many probability density functions, mathematics must determine the area under the curve. When we utilize formulae to calculate the area in this textbook, we employ integral calculus approaches to discover the formulas. We shall not utilize calculus in this textbook since most students in this class have never studied it.

There are a lot of different types of continuous probability distributions. When modeling probability using a continuous probability distribution, the distribution is chosen to best represent and match the specific circumstance.

We'll look at the uniform, exponential, and normal distribution in this and the next chapters. These distributions are shown in the graphs below.

Probability Density Functions

The first step is to create a continuous probability density function. The function notation f is used (x). It's possible that intermediate algebra was your first formal exposure to functions. The functions we explore in the study of probability are unique. The area between f(x) and the x-axis is equivalent to a probability; thus, we define the function f(x). The maximum area is one since the greatest probability is one. Probability = Area for continuous probability distributions [38].

The following graphic depicts the probability density function for a continuous random variable (Fig. **4**):

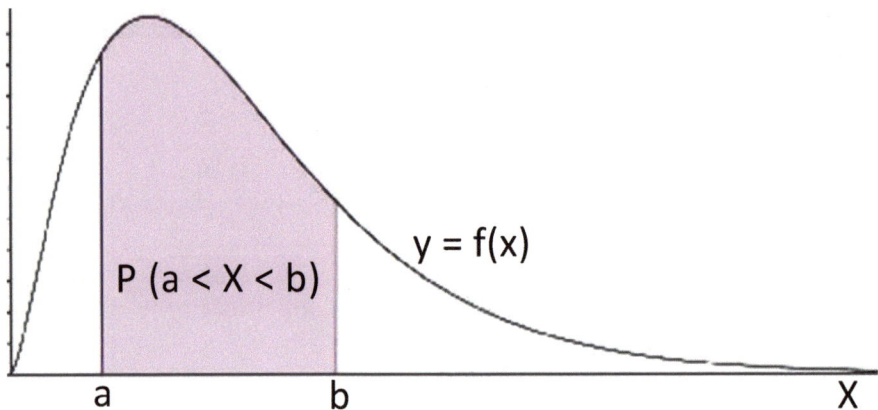

Fig. (4). Probability density function [38].

If a result, X, falls within the range of values a and b, the probability that it falls within that range is indicated by the shaded area under the curve, which can be obtained by determining the integral of the pdf over the specified interval. To put it another way [38].

$$
\begin{array}{cccccc}
11 & 12 & 13 & 14 & 15 & 16 \\
21 & 22 & 23 & 24 & 25 & 26 \\
31 & 32 & 33 & 34 & 35 & 36 \\
41 & 42 & 43 & 44 & 45 & 46 \\
51 & 52 & 53 & 54 & 55 & 56 \\
61 & 62 & 63 & 64 & 65 & 66
\end{array}
$$

It should be noted that if a were negative infinity and b were positive infinity [38].

x	2	3	4	5	6	7	8	9	10	11	12
$P(x)$	$\frac{1}{36}$	$\frac{2}{36}$	$\frac{3}{36}$	$\frac{4}{36}$	$\frac{5}{36}$	$\frac{6}{36}$	$\frac{5}{36}$	$\frac{4}{36}$	$\frac{3}{36}$	$\frac{2}{36}$	$\frac{1}{36}$

Since there can be a 100-point difference that X will come inside this interval if it is a real number, the probability across the full pdf must be equal to one.

Cumulative Distribution Functions

It is advantageous to have a probability density, but it also implies that we will have to solve an integral every time we wish to compute a probability. To prevent this undesirable outcome, we will be using a standard known as a cumulative distribution function (CDF). The CDF is a function that takes a number as an input and returns the probability that a random variable will have a quantity less than that number as a result. It has the pleasant property that, if we have a CDF for a random variable, we don't need to integrate to answer probability questions! The Cumulative Distribution Function, often known as the CDF, is calculated for a continuous random variable F(a), is [38].

$$
P(X \geq 9) = P(9) + P(10) + P(11) + P(12) = \frac{4}{36} + \frac{3}{36} + \frac{2}{36} + \frac{1}{36} = \frac{10}{36} = 0.2\bar{7}
$$

Why is the CDF likely that a random variable takes on a value less than the input value rather than a value larger than the input value instead of vice versa? It is a question of etiquette. However, it is a helpful norm to follow. Knowing the CDF and using the fact that the integral throughout the range negative infinity to positive infinity equals 1 may answer the vast majority of probability issues with relative ease. Some examples of how you can solve probability problems simply by utilizing a CDF include [38]:

$$P\,(X \text{ is even}) \;=\; P\,(2)+P\,(4)+P\,(6)+P\,(8)+P\,(10)+P\,(12)$$
$$=\; \tfrac{1}{36}+\tfrac{3}{36}+\tfrac{5}{36}+\tfrac{5}{36}+\tfrac{3}{36}+\tfrac{1}{36} \;=\; \tfrac{18}{36}=0.5$$

Examples of Probability Distribution of Continuous Random Variable

Example # 1

To catch the next bus to his destination, a man shows up at a bus stop completely at random (*i.e.,* without regard for the time of day or the scheduled service). Given that buses run every 30 minutes on a constant schedule, the next bus will arrive at any moment during the next 30 minutes, with an equally distributed possibility of arriving at any location (a uniform distribution). It is necessary to assess whether a bus will come within the next 10 minutes.

Solution

It is the x-axis everywhere else, and a horizontal line above the range from 0 to 30 represents the graph of the density function. As the entire area under the curve must equal one, the height of the horizontal line must be one-third of the total portion under the curve. See the chart under "Probability of Waiting for a Bus for No More Than 10 Minutes." The probability desired is P(0≤X≤10). This probability is defined as the size of the rectangular region bordered above by the horizontal line f(x)=1/30, bordered below by the x-axis, bordered on the left by the vertical line at 0 (the y-axis), and bordered on the right by the vertical line at 10. This is the shaded zone in Fig. (5) "Probability of Waiting At Most 10 Minutes for a Bus". The rectangle area is equal to the product of the rectangle's base and the height of the rectangle, which is 10(1/30)=13. As a result, P(0≤X≤10)=1/3.

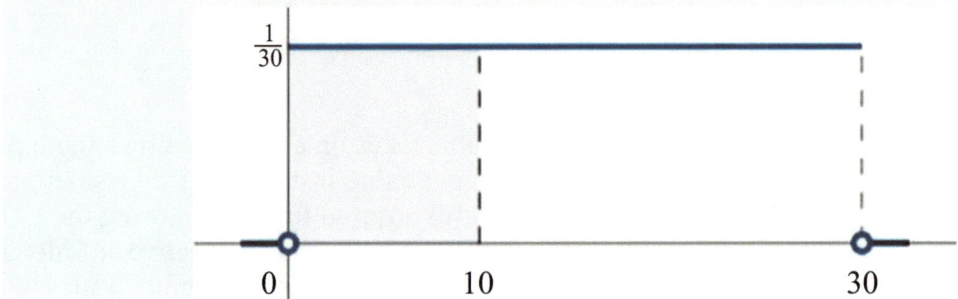

Fig. (5). Probability of Waiting At Most 10 Minutes for a Bus.

The Normal Distribution

The essential distributions are normal, which is a continuous distribution. It's frequently used, but much more so when it's misused. It has a bell-shaped graph (Fig. **6**). The bell curve may be found in practically every field. Psychology, business, economics, the sciences, nursing, and, of course, mathematics are among them. Some of your lecturers may use the normal distribution to decide your grade. The majority of IQ scores follow a normal distribution. Real-estate values often follow a normal distribution. Although the normal distribution is important, it cannot be applied to everything in the actual world [37, 39].

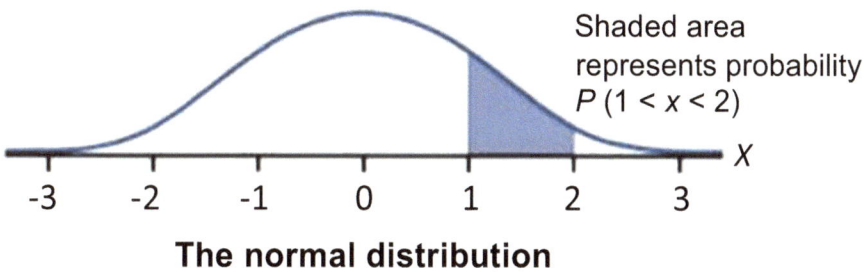

Fig. (6). Normal distribution [37].

The mean (μ) and the standard deviation (σ) are the two parameters (two descriptive numerical measurements) of a normal distribution (Fig. **7**). We identify X as having a normal distribution with mean (μ) and standard deviation (σ) if it is a quantity to be measured [39].

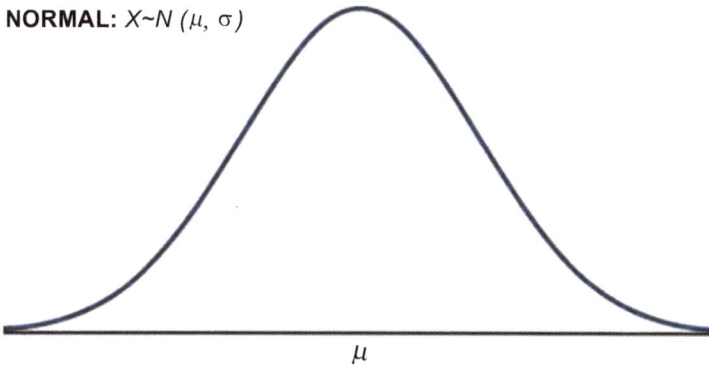

Fig. (7). Normal Distribution [37].

One of the most often observed continuous random variables has a probability distribution that is shaped like a bell (or bell curve) [37].

$$f(x) = \frac{1}{\sigma\sqrt{2\pi}} e^{-1/2}[(x - \mu)/\sigma]^2$$

The normal distribution is significant in statistical inference, and the symbol n represents it. Furthermore, many business occurrences create random variables with probability distributions that are rather close to the normal distribution in terms of their fit to the normal distribution. Example: The monthly rate of return for a specific stock is nearly equal to that of a normal random variable, and the probability distribution for a corporation's weekly sales may be about equal to that of a normal random variable. The normal distribution may also be an appropriate model for distributing employment aptitude exam results if the scores are distributed uniformly.

Understanding Normal Distribution

When it comes to technical stock market analysis and other sorts of statistical studies, the normal distribution is the most often accepted type of distribution to be true. The mean and standard deviation of the standard normal distribution are the two parameters of the distribution. Between one standard deviation (SD) of the mean for a normal distribution, 68 percent of the observations are within two, 95 percent are within three, and 99.7 percent are within four. The Central Limit Theorem serves as the inspiration for the normal distribution model. As stated in the title, this theory says that averages determined from independent, identically distributed random variables have distributions that are essentially normal, regardless of the kind of distribution from which the variables have been collected (provided it has finite variance). Occasionally, the normal distribution is mistaken for the symmetrical distribution. Asymmetrical distribution can have two mirror copies when the data is divided by a dividing line; however, the real data might have two humps or a sequence of hills in addition to the bell curve that signals a normal distribution [39].

Kurtosis and Skewness

Real-world data is seldom, if ever, distributed in a perfectly typical manner. A particular distribution's skewness and kurtosis coefficients indicate how dissimilar it is from a normal distribution in terms of its shape. The skewness of a distribution is an estimate of the symmetry of the distribution. The normal distribution is symmetric and has a skewness of zero, indicating perfectly symmetric. It is implied that the left tail of the distribution is longer than the right tail if the skewness is less than zero or if the distribution has negative skewness; conversely, a positive skewness means that the right tail is longer than the left tail.

The kurtosis measures the thickness of a distribution's tail ends compared to the thickness of the tails of the normal distribution. Tail data from distributions with high kurtosis outnumber the tails of the normal distribution by a factor of two (*e.g.*, five or more standard deviations from the mean). If you have a low kurtosis distribution, you will get tail data that is less severe than the tails of the normal distribution, which is good news. This means that the normal distribution has neither fat nor thin tails. The kurtosis of three implies that the normal distribution does not have fat or thin tails. To put it another way, when you compare an observational distribution to the normal distribution and find that it has a kurtosis bigger than three, you may say that it has heavy tails. Compared to the normal distribution, a distribution is said to have thin tails if the kurtosis is smaller than three, indicating that the distribution has thin tails.

Central Limit Theorem

The central limit theorem (abbreviated as clt) is one of statistics' most powerful and valuable concepts. The theorem has two different versions, both of which take finite samples of size n from a population with a known mean and standard deviation. According to the first option, if we gather samples of size n with a "big enough n," compute each sample's mean, then generate a histogram of those means, the final histogram will resemble a regular bell shape. According to the second option, if we gather samples of size n that are "big enough," compute the total of each sample, and generate a histogram, the final histogram will have a typical bell shape again [40].

The sample size, n, that is necessary to be "big enough," is determined by the population from which the samples are obtained (the sample size should be at least 30, or the data should come from a normal distribution). If the original population is ordinary, additional observations will be required to make the sample means or sums normal. Replacement sampling is used.

Estimating the central limit theorem's significance in statistical theory is impossible. Knowing that data acts predictably, even if its distribution is not normal, is a valuable tool.

The Central Limit Theorem (CLT) lays forth the requirements that must be met in order for the sample mean to converge to a normal distribution as the sample size grows.

Sample Mean

We must first establish the sample mean because Central Limit Theorems are concerned with it.

Let [X_n] be a random variable sequence.

The sample mean of the first n terms of the sequence will be denoted by X-$_n$ [40]

x	-2	1	2	3.5
$P(x)$	0.21	0.34	0.24	0.21

We expand the sample size n by adding more observations X_i to the sample mean. The sample mean is a random variable since it is a sum of random variables.

Convergence to Normal Distribution

We can learn more about the Central Limit Theorem by looking at what happens to a sample's mean distribution as we increase the sample size. Remember that if the criteria of a Law of Large Numbers are met, the sample means will descend in probability to the expected value of the observations. The mean of the observations will converge in probability to the mean of the observations [40].

$$\mu = \Sigma x\, P(x)$$
$$= (-2) \cdot 0.21 + (1) \cdot 0.34 + (2) \cdot 0.24 + (3.5) \cdot 0.21 = 1.135$$

In order to calculate the Central Limit Theorem, we must first normalize the sample mean, which means subtracting from it the anticipated value and dividing the result by the standard deviation. Then, as the sample size increases, we examine the behavior of the distribution's distribution. In practice, this means that the standardized sample means probability distribution approaches that of a normal distribution.

x	299	199	99	-1
$P(x)$	0.001	0.001	0.001	0.997

The Standard Normal Distribution

Classified as the z-distribution, the standard normal distribution is a specific normal distribution in which the mean is zero and the standard deviation is one. The standard normal distribution (Fig. **8**) is defined as follows:

Any normal distribution may be standardized by turning its values into Z-scores, which further be standardized. The Z-scores indicate how many standard deviations each value is apart from the mean of the distribution [41].

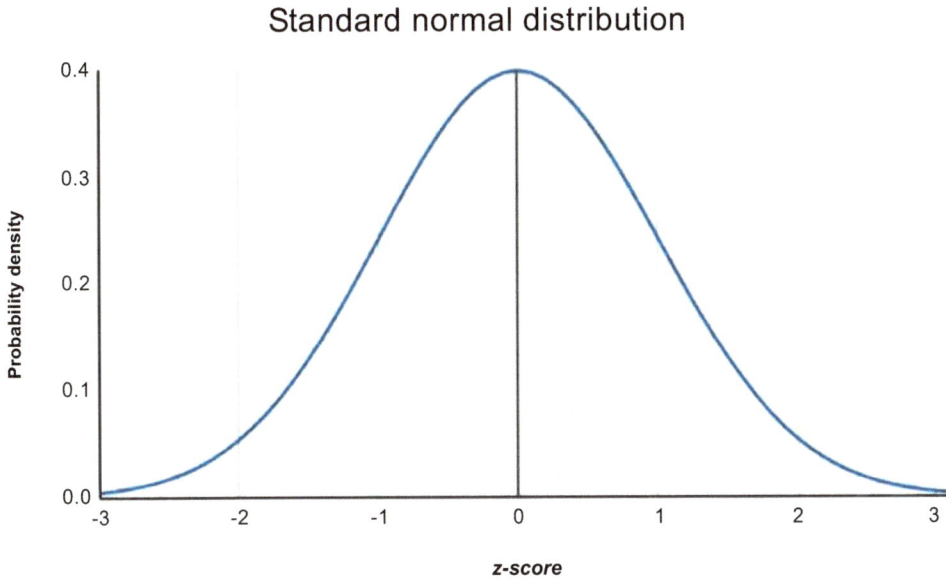

Fig. (8). Standard normal distribution [41].

You may use the z-distribution to compute the likelihood of specific values and compare various data sets after converting a normal distribution to a z-distribution.

The Standard Normal Distribution Vs. Normal Distribution

All normal distributions, including the standard normal distribution, are identical, and functionalists believe a bell-shaped curve is a case with the standard normal distribution. On the other hand, a normal distribution may have whatever mean and standard deviation values are desired. The mean and standard deviation of a standard normal distribution is always set in the same manner [41].

Every normal distribution is a variant of the ordinary, normal distribution that has been stretched or compressed and shifted horizontally to the right or left of its origin [41].

The mean dictates where the curve's center of gravity is located. Expanding the mean shifts the curve to the right, while lowering the mean shifts the curve to the left.

The standard deviation measures how much the curve extends or squeezes. A very tiny standard deviation results in a narrow curve, whereas an extremely high Standard Deviation results in a broad curve (Fig. **9**) [41].

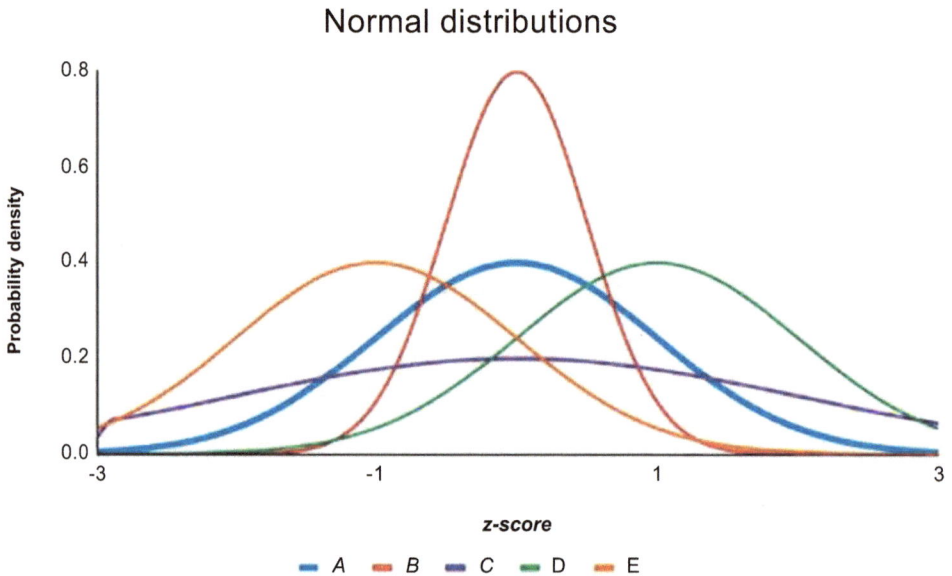

Fig. (9). Standard Distribution *vs* Normal Distribution [41].

Standardizing Normal Distribution

When you standardize a normal distribution, the mean is reduced to zero, and the standard deviation is increased to one. This makes it simple to compute the chance of specific values appearing in your distribution or to compare data sets with various means and standard deviations without having to resort to complex calculations.

In the normal distribution, data points are referred to as x, but in the z-distribution, data points are referred to as Z or Z-scores. Unified standard score that indicates how many standard deviations an individual value (x) is from the mean: A Z-score may be defined as [41].

- In other words, if your x-value is bigger than the mean, your Z-score is positive.
- If your Z-score is negative, it indicates that your x-value is smaller than the mean.
- A Z-score of zero indicates that your x-value is the same as the mean of the group.

When you convert a normal distribution into the standard normal distribution, you may do the following:

- Comparison of results between distributions with various means and standard deviations is a good place to start.
- Normalize scores to make statistically sound decisions (*e.g.*, grading on a curve).
- Determine the chance of observations in a distribution falling above or below a certain value using a probability distribution.
- Calculate the likelihood that a sample mean varies considerably from a known population mean.

How to Calculate Z-score

Standardizing a normal distribution value involves converting the individual value into a Z-score, which is defined as follows [41]:

- Minus your individual value from the mean of the group.
- Split the difference and divide it by the standard deviation.

$$P\left(W\right) = P\left(299\right) + P\left(199\right) + P\left(99\right) = 0.001 + 0.001 + 0.001 = 0.003$$

Example of Finding Z -score

Students enrolled in a new test preparation course provide you with their SAT results. According to the statistics, the mean score (M) is 1150, and the standard deviation (SD) is 150, indicating a normal distribution. You want to know the chance that the SAT scores in your sample will be higher than 1380. [6]

To standardize your data, you must first get the Z-score for 1380. The Z-score shows you how many standard deviations 1380 is from the mean in terms of standard deviations.

<u>Solution</u>

Subtract from the x value to get the y value.

$$E\left(X\right) = 299 \cdot 0.001 + 199 \cdot 0.001 + 99 \cdot 0.001 + \left(-1\right) \cdot 0.997 = -0.4$$

Split the difference by the standard deviation

x	195	$-199{,}805$
$P(x)$	0.9997	0.0003

The Z-score with a value of 1380 is 1.53. That is, the value of 1380 is 1.53 standard deviations away from the mean of your population.

To Find Probability using The Normal Standard Distribution

As the standard normal distribution is a probability distribution, the area under the curve between two points indicates the likelihood of a variable taking on a range of values between the two locations. The whole area under the curve equals one or one hundred percent.

Every Z-score has a corresponding p-value, which gives you the likelihood that all values below or above that Z-score will occur in a given period. In other words, it's the area under the curve that's either left or right of that Z-score (Fig. **10**).

Curve	Position or shape (relative to standard normal distribution)
A ($M = 0$, $SD= 1$)	Standard normal distribution
B ($M = 0$, $SD= 0.5$)	Squeezed, because $SD < 1$
C ($M = 0$, $SD= 2$)	Stretched, because $SD > 1$
D ($M = 1$, $SD= 1$)	Shifted right, because $M > 0$
E ($M = -1$, $SD= 1$)	Shifted left, because $M < 0$

Fig. (10). "Z-score" [41].

P values and Z-Tests

In a Z-test, the Z-score is the test statistic that is employed. For example, to compare the means of two groups or to compare the mean of a group to a predetermined value, the Z-test can be employed. On the other hand, its null hypothesis asserts that there is no difference between groups. The p-value, which is the area under the curve immediately to the right of the Z-score, indicates the chance of your observation happening if the null hypothesis is correct. Generally speaking, a p-value of 0.05 or below implies that your findings are unlikely to have occurred by chance; it suggests the presence of a statistically significant impact in your study. Easily determine the probability of an event occurring based on the normal distribution of the data using the Z-score of a value in the normal distribution [41].

How to Use Z-Table

Once you have a Z-score, you may look up the likelihood associated with that score in a z-table (Table **1**).

Z-tables are used to report the area under the curve (AUC) for every z-value between -4 and 4 at 0.01-unit intervals between -4 and 4 [41].

The z-table may be shown in a number of different ways. We're going to utilize a piece of the cumulative table in this case. This table provides you with the overall area under the curve up to a specific Z-score; the area under the curve is comparable to the chance of values below that Z-score happening in the given time period [41].

Z-tables are organized in columns, each including the Z-score to the first decimal place. The second decimal place is shown in the first row of the table [41].

To calculate the equivalent area under the curve (probability) for a Z-score, do the following calculations:

- Find the row that contains the first two numbers of your Z-score and go down to it.
- Aim for the column with the third digit corresponding to your Z-score.
- The value that lies at the junction of the row and column from the previous stages is found here.

Example: Using Z distribution to Find Probability

We've calculated that an SAT score of 1380 has a Z-score of 1.53. With the whole z-table in hand, the p-value for a Z-score of 1.53 is 0.937 for the condition.

Table 1. Z-table

z	+ 0.00	+ 0.01	+ 0.02	+ 0.03	+ 0.04	+ 0.05	+ 0.06	+ 0.07	+ 0.08	+ 0.09
0.0	0.50000	0.50399	0.50798	0.51197	0.51595	0.51994	0.52392	0.52790	0.53188	0.53586
0.1	0.53983	0.54380	0.54776	0.55172	0.55567	0.55962	0.56360	0.56749	0.57142	0.57535
0.2	0.57926	0.58317	0.58706	0.59095	0.59483	0.59871	0.60257	0.60642	0.61026	0.61409
0.3	0.61791	0.62172	0.62552	0.62930	0.63307	0.63683	0.64058	0.64431	0.64803	0.65173
0.4	0.65542	0.65910	0.66276	0.66640	0.67003	0.67364	0.67724	0.68082	0.68439	0.68793
0.5	0.69146	0.69497	0.69847	0.70194	0.70540	0.70884	0.71226	0.71566	0.71904	0.72240
0.6	0.72575	0.72907	0.73237	0.73565	0.73891	0.74215	0.74537	0.74857	0.75175	0.75490
0.7	0.75804	0.76115	0.76424	0.76730	0.77035	0.77337	0.77637	0.77935	0.78230	0.78524
0.8	0.78814	0.79103	0.79389	0.79673	0.79955	0.80234	0.80511	0.80785	0.81057	0.81327
0.9	0.81594	0.81859	0.82121	0.82381	0.82639	0.82894	0.83147	0.83398	0.83646	0.83891

This is the likelihood of receiving an SAT score of 1380 or below (93.7 percent), which is shown by the area under the curve to the left of the shaded region on the graph (Fig. **11**).

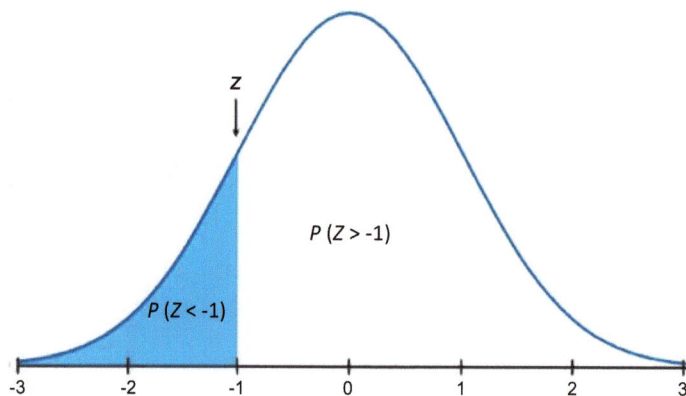

Fig. (11). Using Z distribution to Find Probability [42].

To determine the shaded area, subtract 0.937 from one, which is the total portion under the curve, to get the shaded area.

$$E\left(X\right) = \Sigma x\, P\left(x\right) = 195 \cdot 0.9997 + \left(-199{,}805\right) \cdot 0.0003 = 135$$

Areas of Tails of Distribution

In Fig. (**12**) "Right and Left Tails of a Distribution," the left cut of a curve portion Y=F(X) of a continuous random integer X that a value x has cut off* is the area beneath the curve that is to the left of x*, as illustrated by the shading (a). According to the coloring in Fig. (**12**) "Right and Left Tails of a Distribution," the right tail cut off by x* is defined identically to the left tail cut off by x* (b).

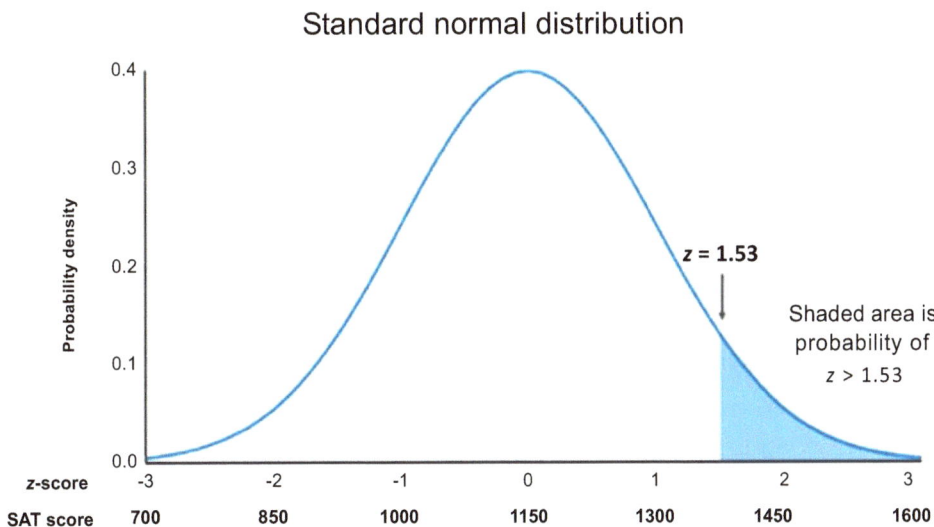

Fig. (12). Right and Left Tails of a Distribution [42].

Tails of Standard Normal Distribution

It is crucial to be able to answer problems of the kind shown in Fig (**12**). at certain points in life. Our attention is focused on a particular area of interest. In this example, the area is 0.0125 of the shaded region in the picture. We want to determine the value z* of the variable Z, which causes it. This is the polar opposite of the kind of difficulties that have been faced so far. The problem is that, instead of knowing the value z* of Z and trying to determine a matching area, we know the area and wish to know the value z*. Using the language of the description just above, we want to discover the value z* that dissects a left tail of area 0.0125 in the ordinary, normal distribution, which is what we're looking for in this situation [42].

The concept behind addressing such an issue is rather straightforward, albeit its execution may be a little challenging at times. Overall, one reads the cumulative probability table for Z in reverse (Fig. **13**), finding the appropriate region inside the table's interior and reading off the value of Z from its margins to summarize.

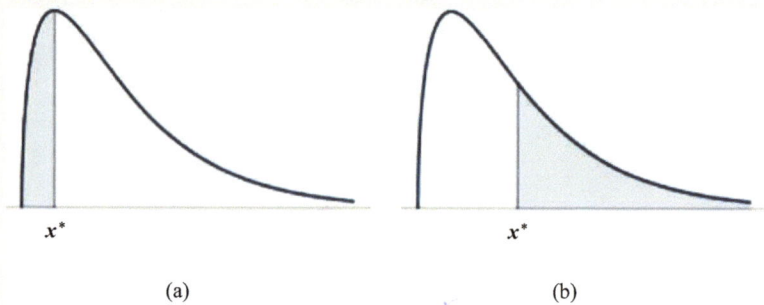

Right and left Tails of a Distribution

(a)   (b)

Fig. (13). Z value [41].

CONCLUSION

In conclusion, visualizing data using distributions is an important tool for researchers and data scientists to use to better show trends in dataset. Both standard and normal datasets and distributions help to show the behavior of variables as well as to assist in discerning probabilities and outcomes. By using these tools, researchers can better understand the behaviors of a dataset as well as provide valuable information about trends and event occurrences. It is important to note that these are not the only methods of showing data regarding independent trials and timed events, but they are the most commonly utilized.

CHAPTER 6

Sampling Distributions

Abstract: Sampling of a distribution and as well as how to treat the sampling distribution of the sample mean is an important topic for understanding statistics. The central limit theorem, when applied to this type of dataset can allow researchers to make predictions about future datasets as well as to better understand current datasets. Normally distributed populations are important to correctly treating the data occurring within the dataset.

Keywords: Bell-curve, Central limit theorem, Normal distribution.

INTRODUCTION

The mean and standard deviation from the sample mean are important information from the dataset that helps researchers to understand the distribution of the data as well as its shape and values. A small sample proportion will show the errors as well as the distance from mean to assist researchers in better understanding the character of their data by allowing researchers to drill down into the nuances of the data.

A sampling distribution is a probability distribution of a statistic derived by randomly selecting a population sample. A finite-sample distribution, also known as a finite-sample distribution, depicts the frequency distribution of how far apart distinct events will be for a given population (Fig. **1**) [43].

The sampling distribution is influenced by several variables, including the statistic, sample size, sampling method, and general population. It is used to determine statistics for a given sample, such as means, ranges, variances, and standard deviations (SD) [43].

A statistic is a number calculated from a sample, such as a sample mean or standard deviation. Every statistic is a random variable since a sample is random: it varies from sample to sample in ways that can't be predicted confidently [44]. It has a mean, a standard deviation (SD), and a probability distribution as a random variable. The sampling distribution of a statistic is the probability distribution of that statistic. Sample statistics are often performed to estimate the corresponding population parameters rather than their goal [44].

Sampling Distribution

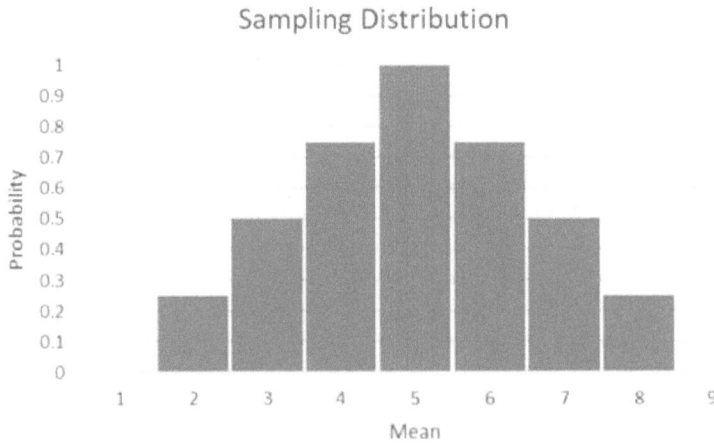

Fig. (1). Sampling Distribution [43].

The mean, standard deviation (SD), and sampling distribution of a sample statistic are introduced in this chapter, with a focus on the sample mean $\bar{x}$ [44].

THE MEAN AND STANDARD DEVIATION (SD) OF THE SAMPLE MEAN

Let's say we want to calculate a population's mean μ. In actuality, we would usually collect one sample. Consider the case where we collect sample after sample of the same size n and calculate the sample mean $\bar{x}$. For each. Each time, we'll most likely obtain a different $\bar{x}$ value [44]. The sample mean $\bar{x}$ is a random variable, meaning that it fluctuates from sample to sample in unpredictable ways. When we think of the sample mean as a random variable, we'll write $\bar{X}$ for the values it takes. The random variable $\bar{X}$, has a mean denoted as μ x and a standard deviation (SD) indicated by the letters σ x. Here's an example where the population is so tiny, and the sample size is so small that we can write down every sample [44].

Examples

Example 1

Four rowers, weighing 152, 156, 160, and 164 pounds, make up a rowing team. Calculate the sample mean for each available random sample with size two replacement. Use these to calculate the sample mean $\bar{X}$, probability distribution, mean, and standard deviation (SD) [44].

Solution

The following table lists all potential samples with size two replacement, as well as the mean of each [44]:

Sample	Mean		Sample	Mean		Sample	Mean		Sample	Mean
152, 152	152	-	156, 152	154	-	160, 152	156	-	164, 152	158
152, 156	154	-	156, 156	156	-	160, 156	158	-	164, 156	160
152, 160	156	-	156, 160	158	-	160, 160	160	-	164, 160	162
152, 164	158	-	156, 164	160	-	160, 164	162	-	164, 164	164

The sample mean $\bar{X}$ has seven different values, as shown in the table. The number $\bar{x}=152$, like the value $\bar{x}=164$, occurs just once (the rower weighing 152 pounds must be picked both times), while the other values occur several times and are therefore more likely to be detected than 152 and 164. Because the 16 samples are all equally probable, we can count to get the sample mean's probability distribution [44]:

$\bar{x}$	152	154	156	158	160	162	164
$P(\bar{x})$	1/16	2/16	3/16	4/16	3/16	2/16	1/16

We use the formulae for the mean, and standard deviation of a discrete random variable from Section 4.3.1 "The Mean and Standard Deviation of a Discrete Random Variable" in Chapter 4 "Discrete Random Variables", get a result for $\mu\bar{X}$ [44].

$\mu\bar{X}.= \Sigma x- P(-)=152(116)+154(216\}+156(316)+158(416)+160(316\}+162(216)+164(116)=158$

For $\sigma X-$ we first compute $\Sigma x-2P(x-)$:

$1522(116)+1542(216)+1562(316)+1582(416)+1602(316)+1622(216)+1642(116)$

which is 24,974, so that

$\sigma X =\Sigma x-2P (x-) - \mu x-2=24,974-1582=10$

In the example, the mean and standard deviation (SD) of the population {152,156,160,164} are $\mu = 158$ and $\sigma= 20$ respectively. The mean of the sample mean $\bar{X}$ that we just calculated is identical to the population mean. We just calculated the standard deviation (SD) of the sample mean $\bar{X}$, which is the population standard deviation (SD) divided by the square root of the sample size:

10 = 20/ 2. These connections are not coincidental; they are instances of the following formulae.

Suppose random samples of size n are drawn from a population with mean μ and standard deviation σ. The mean μ_X and standard deviation σx of the sample mean $\overline{X}$ satisfy [44]

$\mu_X = \mu$ and $\sigma x = \sigma / n$

According to the first formula, if we were to take every conceivable sample from the population and calculate the appropriate sample mean, the numbers would be centered around the number we want to estimate, the population means μ [44].

The second formula quantifies the link by stating that averages calculated from sampling vary less than individual population values [44].

Example 2

The mean and standard deviation (SD) of all automobiles registered in a given state's tax value are $\mu=\$13,525$ and $\sigma=\$4,180$, respectively. Assume that 100 random samples are taken from the population of automobiles. What is the sample mean μx^{-} mean σx^{-} and standard deviation (SD) of the sample mean $\overline{X}$?

Solution:

Because n = 100, the calculations result in [44]:

$\mu_X = \mu = \$13,525$ and $\sigma x = \sigma /n = \$4180 / 100 = \$418$

The Sampling Distribution of the Sample Mean

In statistics, the sampling distribution is a probability distribution based on a large number of small sample sizes drawn from a specific population to provide a statistical result [45].

Consider the following example. The delivery of a big tank of fish from a hatchery to the lake is now taking place. The average length of the fish in the aquarium is what we're looking for. As opposed to taking measurements on all of the fish, we randomly picked twenty fish and used the sample mean to determine the mean of the whole population [45].

The sample mean of the twenty fish is denoted by the letter $\overline{x}_1$. Consider the following scenario: we get a different sample of size twenty from the same hatchery. The sample mean is denoted by the symbol $\overline{x}_2$. Would $\overline{x}_1$ equal $\overline{x}_2$ in this case? It is not always the case. Imagine if we collected another sample and calculated the mean. Consider the following scenario: you take 1000 random samples of size twenty and record the means of all of the samples. We might generate a histogram by taking the 1000 sample averages and dividing them by 100. It would provide us with a visual representation of the distribution of the sample means. The sampling distribution of the sample mean is defined as the distribution of all of the sample means combined [45].

To estimate a certain population parameter of interest, we may discover the sampling distribution of any sample statistic. We will be looking at the sampling distributions for the sample mean, which is denoted by x‾ And the sample proportion, which is denoted by p^ [45].

Starting with a description of the sampling distribution of the sample mean, we go on to apply the central limit theorem. After that, we'll talk about how sample proportions are distributed in sampling. The mean of all sample means (x-bars) is the population mean if a large number of selected samples of a certain size are taken n are data gathered from a large number of possible values for a quantitative variable, in which the population is concentrated mean is μ (mu), and the population standard deviation (SD) is σ (sigma) then the mean of all sample means (x-bars) is the population mean μ (mu) [45].

Theory prescribes the behavior much more accurately than claiming that bigger samples have less spread when it comes to the spread of all sample means. In reality, as shown below, the standard deviation (SD) of all sample means is proportional to the sample size n [45].

The standard deviation of all sample means $\overline{(x)}$ ı is exactly σ /n.

The Central Limit Theorem

For samples of size two chosen from a population of four rowers, we computed the probability distribution of the sample mean. The following is the probability distribution [44]:

$\overline{(x)}$ ı	152	154	156	158	160	162	164
p $\overline{(x)}$ ı	1/16	2/16	3/16	4/16	3/16	2/16	1/16

A histogram for the original population and a histogram for this distribution are shown side by side in the Fig. (**2**). While the population's distribution is uniform, the sampling distribution of the mean has a shape that resembles the well-known bell curve. This phenomenon of the mean sample distribution taking on a bell shape despite not being a bell-shaped population often occurs. Here's an example that's a little more realistic (Fig. **2**).

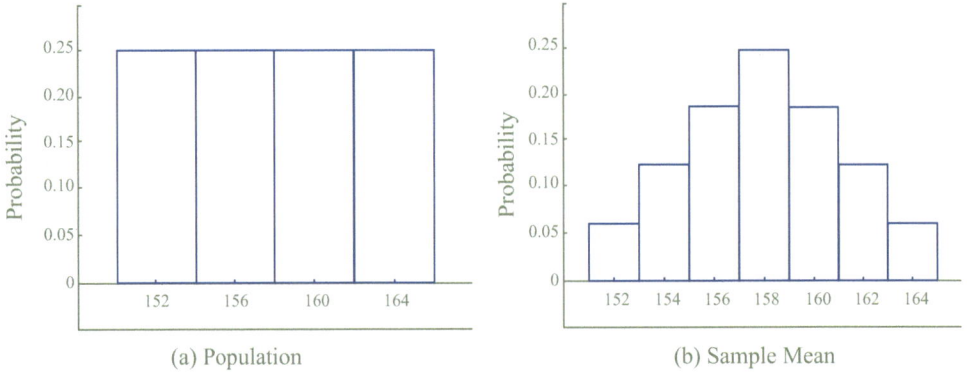

(a) Population   (b) Sample Mean

Fig. (2). Probability [44].

Assume we collect samples of size 1, 5, 10, or 20 from a population made up exclusively of the integers 0 and 1, with half of the population being 0 and the other half being 1, resulting in a population mean of 0.5. The following are the sample distributions:

n=1:

$\bar{(x)}_i$	0	1
$p_{\bar{(x)}_i}$	0.5	0.5

n=5:

$\bar{(x)}_i$	0	0.2	0.4	0.6	0.8	1
$p_{\bar{(x)}_i}$	0.03	0.16	0.31	0.31	0.16	0.03

n=10:

$\bar{(x)}_i$	0	0.1	0.2	0.3	0.4	0.5	0.6	0.7	0.8	0.9	1
$p_{\bar{(x)}_i}$	0.00	0.01	0.04	0.12	0.21	0.25	0.21	0.12	0.04	0.01	0.00

n=20:

$\bar{x}$	0	0.05	0.10	0.15	0.20	0.25	0.30	0.35	0.40	0.45	0.50

$p(\overline{x})$	0.00	0.00	0.00	0.00	0.00	0.01	0.04	0.07	0.12	0.16	0.18

$(\overline{x})$	0.55	0.60	0.65	0.70	0.75	0.80	0.85	0.90	0.95	1
$p(\overline{x})$	0.16	0.12	0.07	0.04	0.01	0.00	0.00	0.00	0.00	0.00

The sampling distribution (Fig. **3**) "Distributions of Sample Mean" of $\overline{X}$ changes in an unusual manner as n increase: the probability on the lower and higher ends diminish, but the probabilities in the center grow in respect to them. The form of the sample distribution would grow smoother and more bell-shaped as n was increased further [44].

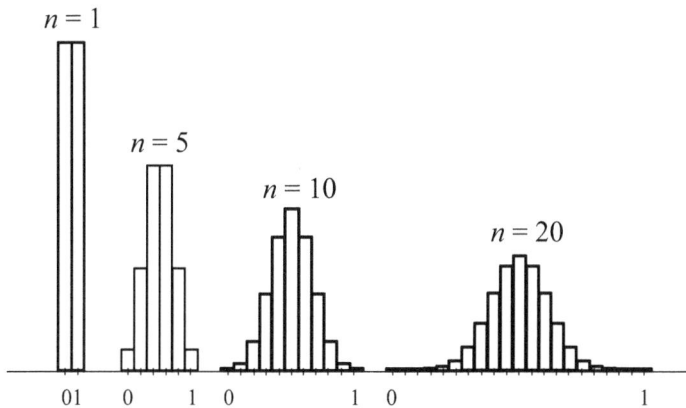

Fig. (**3**). Distributions of Sample Mean [44].

In these cases, the results are independent of the population distributions involved. In principle, any distribution may be used as a starting point, and the sampling distribution of the sample mean will approximate the normal bell-shaped curve as the sample interval grows. The (CLT) Central Limit's Theorem substance is as follows:

For samples of size 30 or more, the sample mean is approximately normally distributed, with mean $\mu_x = \mu$ and standard deviation $\sigma x = \sigma /n$, where n is the sample size. The larger the sample size, the better the approximation [44].

In the Fig. (**4**) titled "Distribution of Populations and Sample Means," the Central Limit Theorem is (CLT) demonstrated for numerous typical population distributions (Fig. **4**).

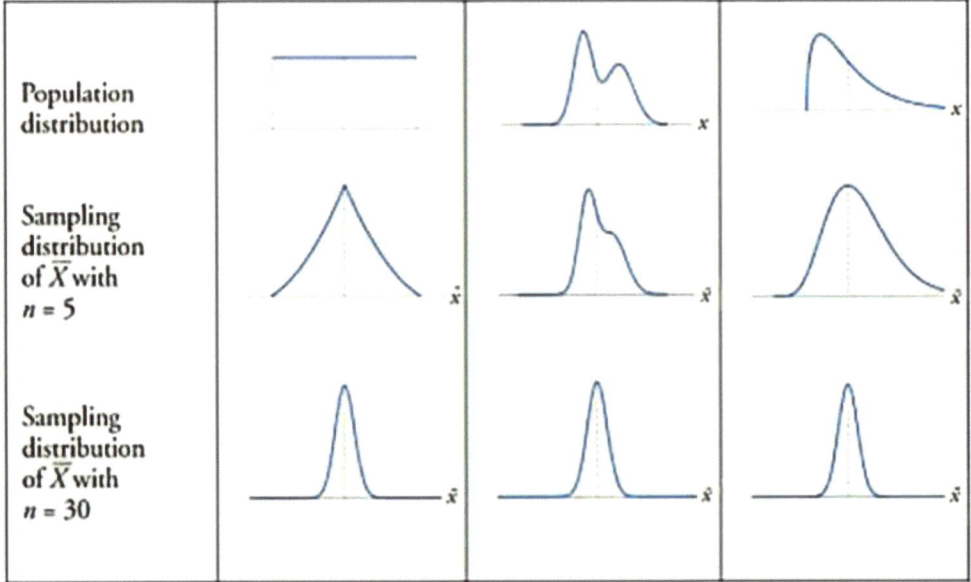

Fig. (4). Distribution of Populations and Sample Means [44].

The population mean is used in the pictures by vertical lines that are dashed in both directions. Regardless of the form of the population distribution, the sampling distribution of the sample mean becomes more bell-shaped as the sample size increases. centered on the population mean, as the sample size increases. After 30 participants, the collection mean's distribution begins to approach that of the normal distribution [44].

As we shall see in the examples, the Central Limit Theorem (CLT) is important because it enables us to make probabilistic judgments about the sample mean, especially concerning its value in contrast to the population mean. However, to correctly apply the finding, we must first recognize that there are two different random variables (and hence two probability distributions) at work:

X measurement of a single element picked at random from the population; the distribution of X is the distribution of the population, with the mean being the population mean μ and standard deviation (SD) σ being the population standard deviation (SD); $\overline{X}$, the mean of measurements in a sample of size n; the distribution of $\overline{X}$ is the sampling distribution, with mean $\mu_{\overline{x}} = \mu$ and standard deviation (SD) $\sigma_{\overline{x}} = \sigma/n$, the mean of measurements in a sample of size n; the distribution of X is the sampling distribution of the population, with mean [44].

Examples

Example 1

For simplicity, we'll $\bar{X}$ use the mean of a sample of size 50 selected from a population with a mean of 112 and a standard deviation (SD) of 40 as our reference point.

a. Calculate the mean and standard deviation (SD) of the variable $\bar{X}$.
b. The likelihood that $\bar{X}$ will take on a value between 110 and 114 is to be determined.
c. The chance that $\bar{X}$ takes on a value larger than 113 is to be determined.

Solution

a. According to the formulae in the preceding section:

$\mu_X = \mu = 112$ and $\sigma x = \sigma / n = 40 / 50 = 5.65685$ [44]

b. The Central Limit Theorem (CLT) is applicable since the sample size is at least 30 people: $\bar{X}$ has a nearly normal distribution. Figuring out probabilities; the only difference is that we make sure to use $\sigma \bar{X}$ and not σ when we standardize:

$P(110 < \bar{X} < 114) = P(110 - \mu_X / \sigma x < Z < 114 - \mu_X / \sigma x)$

$= P(110 - 112 / 5.65685 < Z < 114 - 112 / 5.65685)$

$= P(-0.35 < Z < 0.35) = 0.6368 - 0.3632 = 0.2736$ [44]

c. Similarly,

$P(\bar{X} > 113) = P(Z > 113 - \mu_X / \sigma x)$

$= P(Z > 113 - 112 / 5/65685)$

$= P(Z > 0.18)$

$= 1 - P(Z < 0.18) = 1 - 0.5714 = 0.4286$ [44]

Note that if we had been asked to compute the probability that the value of a single randomly taken object of the population exceeds 113, that is, to compute the number $P(X > 113)$, we would have been unable to do so because we do not know the distribution of X, only that its mean is 112 and its standard deviation (SD) is 40, as we did in the previous example. Instead, we could calculate $P(\bar{X} > 113)$ even if we didn't know the distribution of X, since the Central Limit

Theorem assures that the distribution of X^- is nearly normal [44].

Example 2

The mean and standard deviation (SD) of the total number of grade point averages earned by students at a certain institution are 2.61 and 0.5, respectively. What is the likelihood that a random sample of size 100 is drawn from the population and that the sample mean will fall between 2.51 and 2.71?

Solution [44]

The sample mean X^- has mean $\mu_x = \mu = 2.61$ and standard deviation

$\sigma x = \sigma / n = 0.5/10 = 0.05$, so

$P(2.51 < X^- < 271 = P(2.51 - \mu_x / \sigma x < Z < 2.71 - \mu_x / \sigma x)$

$= P(2.71 - 2.61/0.05 < Z < 2.71 - 2.61/0.05)$

$= P(-2 < Z < 2)$

$= P(Z < 2) - P(Z < -2)$

$= 0.9772 - 0.0228 = 0.9544$

Example 3

With a Small Population, Sample Means are calculated: Weights Made of Pumpkin

Using this illustration, the population is represented by the weight of six pumpkins (in pounds) placed at a carnival game booth where participants must predict the weight of the pumpkins. You are tasked with estimating the average weight of the six pumpkins based on a random sample taken from the population without replacing any pumpkins [45].

Pumpkin	A	B	C	D	E	F
Weight (in pounds)	19	14	5	9	10	17

We can determine the population mean since we know the population weights [45].

$\mu = 19 + 14 + 15 + 9 + 10 + 17 / 6 = 14$ pounds

To show the sampling distribution, let's start by taking all of the feasible n=2 samples from the populations, sampling without replacement. The table below lists all of the available samples and the weights assigned to the selected pumpkins, the sample mean, and the likelihood of receiving each sample. Each sample will have the same chance of getting picked since we are picking at random.

We can put all of the numbers together to make a table with all potential outcomes and their probability [45].

$\overline{x}$	9.5	11.5	12.0	13.0	13.5	14.0	14.5	15.5	16.0	17.0	18.0
Probability	1/15	1/15	2/15	1/15	1/15	1/15	2/15	1/15	1/15	1/15	1/15

When the sample size is 2, the table is the probability table for the sample mean, and the sampling distribution of the sample mean weights of the pumpkins. It's also worth noting that the total probability equals 1. A graph of these numbers could be useful.

As can be seen (Fig. **5**), the probability of the sample mean matching the population mean is just 1 in 15, which is very tiny. (In certain other cases, the sample mean and population mean may never be the same.) Because the sample mean is random, there is a possibility of a mistake when using it to estimate the population mean [45].

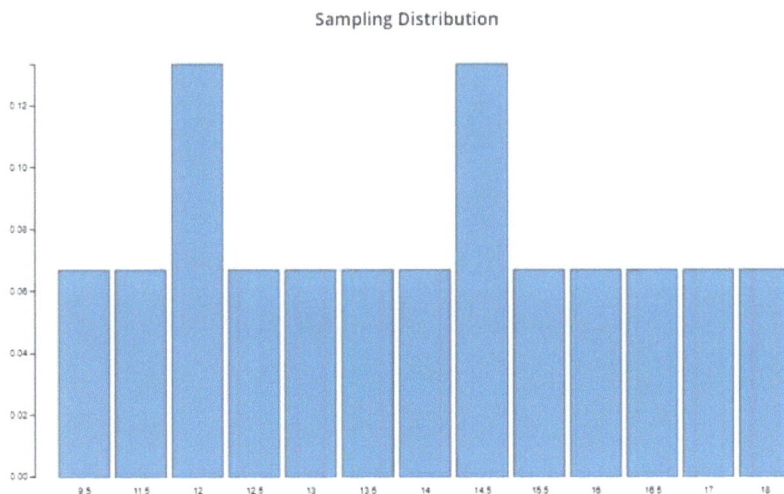

Fig. (5). Probability of sample mean [45].

Now that we know the sampling distribution of the sample mean, we can determine the mean of all the sample means. Put another way, and we may calculate the mean (or anticipated value) of all possible.

The average of the sample means as follows:

$\mu_x = \Sigma \bar{x} f(\bar{x}_i) = 9.5(1/15) + 11.5 (1/15) = 12 (2/15) + 12.5 (1/15) + 13 (1/15) + 13.5 (1/15) + 14 (1/15) + 14. 5 (2/15) + 15$

Although each sample may provide an answer that contains some mistake, the predicted result is perfectly on target: the population mean. To put it another way, if the experiment is repeated several times, the overall average of the sample mean equals the population mean.

Let's try it again, but this time with a sample size of n=5.

Sample	Weights	$\bar{x}$	Probability
A,B,C,D,E	19,14,15,9,10	13.4	1/6
A,B,C,D,F	19,14,15,9,17	14.8	1/6
A,B,C,E,F	19,14,15,10,17	15.0	1/6
A,B,D,E,F	19,14,9,10,17	13.8	1/6
A,C,D,E,F	19,15,9,10,17	14.0	1/6
B,C,D,E,F	14,15,9,10,17	13.0	1/6

The following is the sample distribution:

$\bar{x}$	13.0	13.4	13.8	14.0	14.8	15.0
Probability	1/6	1/6	1/6	1/6	1/6	1/6

The average of the sample means as follows:

$\mu = (1/6) (13 + 13.4 + 13.8 + 14.0 + 14.8 + 15.0) = 14$ pounds

The distribution of sample means for sample sizes of n=2 and n=5 is seen in the dot plots below (Fig. **6**) [45].

Population Mean

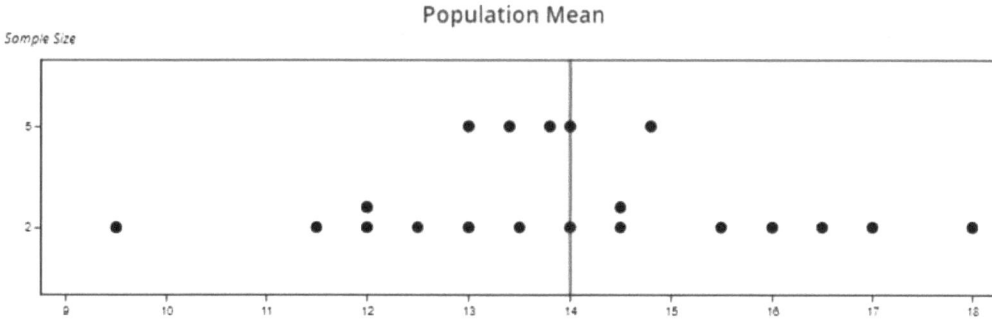

Fig. (6). Population Mean [45].

We can see that using the sample mean to determine the population mean is subject to sampling error. However, with a sample size of n=5, the error is lower than with a sample size of n=2 [45].

Size and Sampling Error

Sample size and inaccuracy in sampling: As the sample grows, the potential sample means cluster closer to the population mean, as seen in the dot plots above. As a result, as the sample size grows, the possibility of sampling error reduces.

What happens if the population isn't so tiny, as in the case of the pumpkin?

Example 4

Using Large Samples to Calculate Sample Means

There are 200 students in an introduction to statistics class taught by a professor. The scores out of 100 points are presented in the histogram (Fig. 7) [45].

The population mean is $\mu = 71.18$ and the population standard deviation is $\sigma = 10.73$ [45].

Normally Distributed Populations

It is stated in the Central Limit Theorem that, no matter how diverse the population's distribution is, as long as the sample is "big," defined as having a sample size of 30 or more, the sample mean will be nearly normally distributed. If the population's mean has a normal distribution, to begin with, then the sample mean will have a normal distribution as well, regardless of the size of the sample.

For samples of any size drawn from a normally distributed population, the sample mean is normally distributed, with mean $\mu_x = \mu$ and standard deviation $\sigma x = \sigma /$n, where n is the sample size [44].

Histogram of Exam Scores

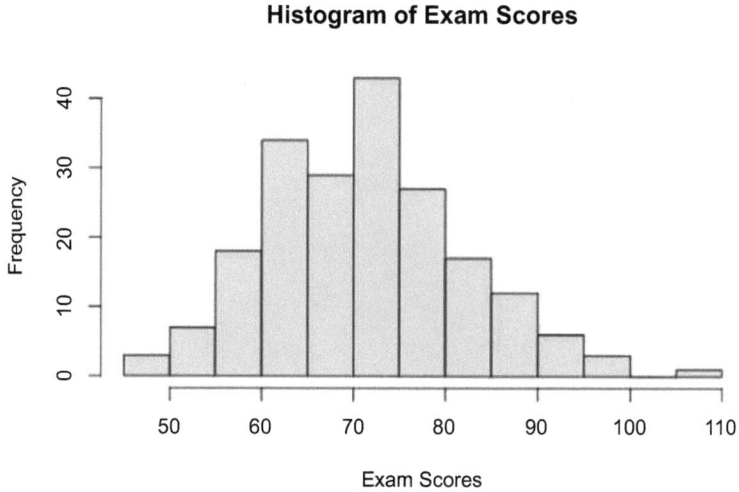

Fig. (7). Histogram [44].

As seen in the (Fig. 8) "Distribution of Sample Means for a Normal Population," the impact of increasing the sample size can be seen.

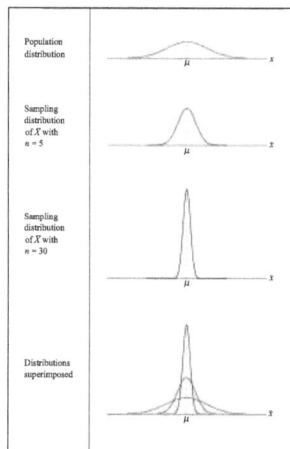

Fig. (8). Distribution of Sample Means for a Normal Population [44].

If the population has a normal distribution with mean μ and standard deviation (SD) σ, the sampling distribution of the sample mean will be normal as well, regardless of sample size [44]. When sampling with replacement is used, or if the population size is larger than the sample size $\bar{x}$, the mean $\mu$and standard deviation (SD) σ/n are used. The standard deviation (SD) of a statistic is referred to as standard error, and because sample average, $\bar{x}$ is a statistic, standard deviation (SD) of $\bar{x}$ is also referred to as standard error of $\bar{x}$. However, you may come across the phrase standard error for the anticipated standard deviation (SD) of $\bar{x}$ in certain literature. The earlier definition is used here, *i.e.*, standard error of x- equals the standard deviation (SD) of $\bar{x}$.

Standard Deviation of $\bar{x}$ (Standard Error) [44]

$$SD(\bar{x}) = SE(\bar{x}) = \sigma/n$$

When we know that the sample mean is Normal or about Normal, we can compute a z-score for the sample mean and calculate the probability for it by using the following formula [2]:

Z-Score of the Sample Mean [44]

$$z = \bar{x} - \mu / \sigma / n$$

Examples

Example 1

Approximately 38,500 miles is the design life of a prototype vehicle tire, with a standard variation of 2,500 miles. A total of five of these tires are being made and tested. Assuming the real population means 38,500 miles and the actual population standard deviation (SD) is 2,500 miles, calculate the likelihood that the sample mean will be less than 36,000 miles under the conditions given in the previous section. Assume that the distribution of tire lifespan is typical for this kind of tire [44].

Solution

For the sake of simplicity, we measure distances in thousands of kilometers. The sample mean $\bar{X}$ has a mean $\mu\bar{X}=38.5$ and a standard deviation (SD) $\sigma\bar{X}=\sigma$ /n=2.5/5=1.11803. The sample standard deviation (SD) X is 2.5/5-=1.11803. Because the population is normally distributed, X is also normally distributed; as a result [44],

$$P(\bar{X} <36) = P(Z < 36 - \mu_x /\sigma_x)$$

$$= P(Z < 36\text{-}38.5/1.11803)$$

$= P(Z < -2.24) = 0.0125$

If the tires operate as intended, there is a 1.25 percent probability that the average of a sample of this size will be as low as it was in the first place.

Example 2

A vehicle battery manufacturer states that their midgrade battery has a mean life of 50 months with a standard variation of 6 months and a standard deviation (SD) of 6 months. Take, for example, a brand whose battery lifetimes are distributed in an almost typical manner.

a. Calculate the likelihood that a randomly chosen battery of this kind would survive fewer than 48 months on the premise that the manufacturer's promises are correct.
b. Following this assumption, calculate the likelihood that the mean of a random sample of 36 such batteries would last fewer than 48 months [44].

Solution

a. Because it is well known that the population has a normal distribution [44],

$P(X < 48) = P(Z < 48 - \mu / \sigma) = P(Z < 48 - 50 / 6)$

$= P(Z < 0.33) = 0.3707$

b. The sample mean has mean $\mu x = \mu = 50$ and standard deviation $\sigma x = \sigma / n = 6 / 36 = 1$. Thus,

$P(X < 48) = P(Z < 48 - \mu x / \sigma x)$

$= P(Z < 48 - 50/1)$

$= P(Z < -2) = 0.0228$

Example 3

For speedboat engines, Ford produces engines with an average power of 220 horsepower (HP) and a standard deviation (SD) of 15 horsepower (HP). You may assume that power distribution follows a normal distribution if you want to be conservative.

After evaluating the engines, Consumer Reports® has decided to contest the company's claim if the sample mean less than 215 horsepower. What is the likelihood that the mean of four engines is less than 215 if they take a sample of four engines?

Solution

We want to find $P(\bar{X} < 215)$.

Since the population follows a normal distribution, we can conclude that $\bar{X}$ has a normal distribution with mean 220 HP (μ =220) and a standard deviation of $\dfrac{\sigma}{\sqrt{n}} = \dfrac{15}{\sqrt{4}} = 7.5$HP.

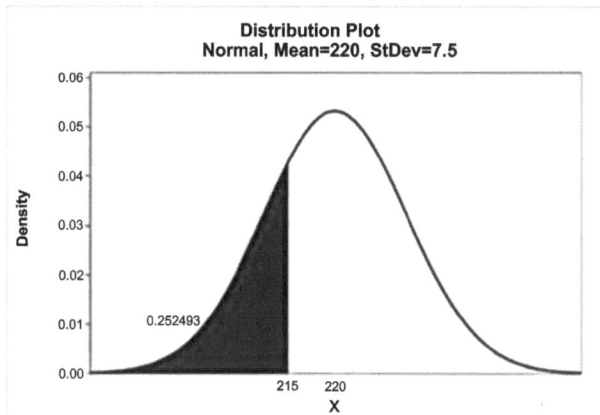

Distribution Plot
Normal, Mean=220, StDev=7.5

$$P(\bar{X} < 215) = P\left(Z < \frac{215 - 220}{7.5}\right) = P(Z < -0.67) \approx 0.2514$$

If Consumer Reports® samples four engines, the probability that the mean is less than 215 HP is 25.14%.

What happens when a sample is drawn from a population that does not follow the normal distribution? The Central Limit Theorem comes into play in this situation.

We should take a moment to dissect what this theorem is trying to communicate since the Central Limit Theorem (CLT) is quite strong!

The Central Limit Theorem may be applied to a sample mean from any distribution, regardless of its shape. We might have a distribution that is biased to the left or the right. Because of the huge number of participants, the sample means will be distributed in a manner that is near to the Normal distribution [45].

When considering the sample size for this course, n>30 is considered a big sample size.

The Sample Proportion

Before we get started, let's go through some of the terminology and notation that is related to proportions:

- The population percentage is denoted by the letter n. It has a predetermined value.
- The size of the random sample is denoted by the letter p.
- The symbol denotes the sample percentage p^. The amount varies depending on the sample [45].

The following example will demonstrate how to calculate the sample distribution for a small population using the sampling distribution formula.

Sample Proportions in a Small Population:

Consider the following scenario: a family with five children is present. Alex (A), Betina (B), Carly (C), Debbie (D), and Edward are the names of the individuals (E). The name of the kid and their favorite color are included in the table below [45].

Name	Alex (A)	Betina (B)	Carly (C)	Debbie (D)	Edward (E)
Color	Green	Blue	Yellow	Purple	Blue

We are interested in the percentage of children in the family that favor the color blue, and from the table, we can see that p=40 of the children prefer the color blue, which is a significant number [45].

Consider the following scenario, which is similar to the pumpkin example from earlier in the lesson: we don't know what percentage of youngsters like blue as their favorite color. To estimate the fraction, we will use resampling techniques. Let us consider n=2 repeated samples obtained without replacing any of them. Here is a list of all the potential samples of size n=2 and their relative odds of having a majority of youngsters who like blue colors (Fig. **9**) [45].

Sampling Distribution of P(Blue)

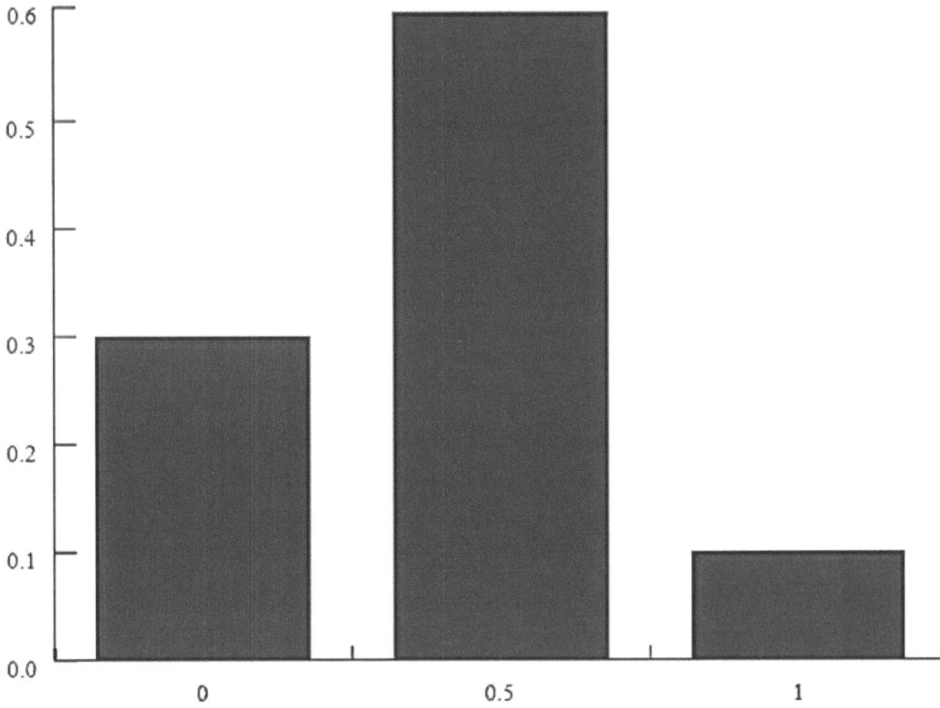

Fig. (9). Sampling Distribution of P [45].

Sample	P (Blue)	Probability
AB	1/2	1/10
AC	0	1/10
AD	0	1/10
AE	1/2	1/10
BC	1/2	1/10
BD	1/2	1/10
BE	1	1/10
CD	0	1/10
CE	1/2	1/10
DE	1/2	1/10

The probability mass function (PMF) is defined as follows [45]:

P (Blue)	0	1/2	1
Probability	3/10	6/10	1/10

In reality, P = P(Blue) =2/5 is the correct percentage. As the PMF illustrates, when the sample size is n=2, it is not feasible to get a sampling percentage equal to the real proportion.

Although it is not discussed in depth here, we might obtain the sampling distribution for higher sample size, such as n=4, if we used a different method. The PMF for the number n=4 is... [45].

P (Blue)	¼	½
Probability	2/5	3/5

The sampling distribution of the sample percentage will have sampling error, just as the sampling distribution of the sample mean would have sampling error [45]. It is also true that the larger the sample size, the narrower the spread of the distribution will be in the final analysis [45].

Often, sampling is done to estimate the percentage of a population with a given attribute, such as the proportion of all faulty goods that come off an assembly line or the proportion of all individuals who visit a retail shop who purchase before leaving. The population percentage is marked by the letter p, whereas the sample proportion is denoted by the letter p^. Thus, if 43 percent of individuals entering a business purchase before leaving, p = 0.43; if in a sample of 200 people entering the store, 78 percent make a purchase, p^=78/200=0.39 [44].

In statistical terms, the sample percentage is a random variable, meaning that it fluctuates from one sample to the next in a manner that cannot be anticipated in advance. When seen as a random variable, it will be denoted by the letter P^. A mean $\mu P^{\wedge}$ and a standard deviation (SD) $\sigma P^{\wedge}$ are assigned to the data set. Here are the formulae for calculating their values [2].

Imagine that random samples of the size nare selected from a population in which the percentage of individuals who possess a characteristic of interest is equal to the number p. The sample proportion $\mu P^{\wedge}$ has a mean and standard deviation (SD) $\sigma P^{\wedge}$ that are both within the acceptable range [44].

$\mu P^{\wedge}$ = p and $\sigma P^{\wedge}$= pq/n

Where q=1-p.

The population proportion $P^\wedge$ has an equivalent in the Central Limit Theorem. Imagine that every member of the population with the desired feature is labeled with a 1, and every element that does not is labeled with a 0. This results in a numerical population made up of zeros and ones. The percentage of the population with the particular characteristic is the proportion of the numerical population that is one; in symbols, the proportion of the population with the special characteristic is the proportion of the numerical population that is one [44].

p = number of 1s / N

However, because the total of all the zeros and ones is just the number of ones, the numerical population's mean is [44]:

$\mu = \Sigma x / N$ = number of 1s / N

As a result, the population proportion p is the same as the mean μ of zeros and ones. Similarly, the sample proportion $p^\wedge$ is equal to the sample mean $\bar{x}$. As a result, the Central Limit Theorem holds for $P^\wedge$. However, the requirement that the sample is big is a bit more involved than just having a size of at least 30 [44].

The Sampling Distribution of the Sample Proportion

For large samples, the sample proportion is approximately normally distributed, with mean $\mu P^\wedge = p$ and standard deviation $\sigma P^\wedge = pq/n$.

A sample is large if the interval $[p - 3\sigma P^\wedge, p + 3\sigma P^\wedge]$ lies wholly within the interval [0, 1] [44].

In reality, p is unknown, and hence neither is $\sigma P^\wedge$. In such a situation, we replace the known amount $p^\wedge$ for p to ensure that the sample is big enough. This entails ensuring that the interval is contained inside the range [0,1]. In the examples, this is shown.

When p = 0.1, a sample size of 15 is too small, while a sample size of 100 is adequate, as shown in the first Fig. (**10**), "Distribution of Sample Proportions." The second Fig. (**11**), "Distribution of Sample Proportions for P," illustrates that a sample size of 15 is suitable for p = 0.5.

Distribution of Sample Proportions

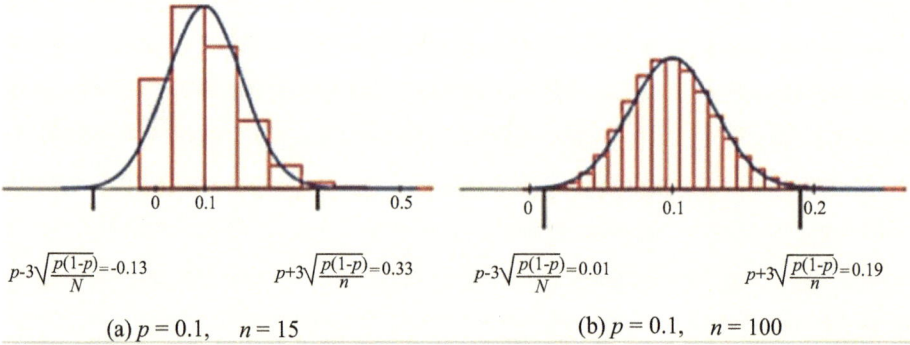

$p-3\sqrt{\frac{p(1-p)}{N}}=-0.13$ $p+3\sqrt{\frac{p(1-p)}{n}}=0.33$ $p-3\sqrt{\frac{p(1-p)}{N}}=0.01$ $p+3\sqrt{\frac{p(1-p)}{n}}=0.19$

(a) $p = 0.1$, $n = 15$ (b) $p = 0.1$, $n = 100$

Fig. (10). Distribution of Sample Proportions [44].

Distribution of Sample Proportions for p = 0.5 and n = 15

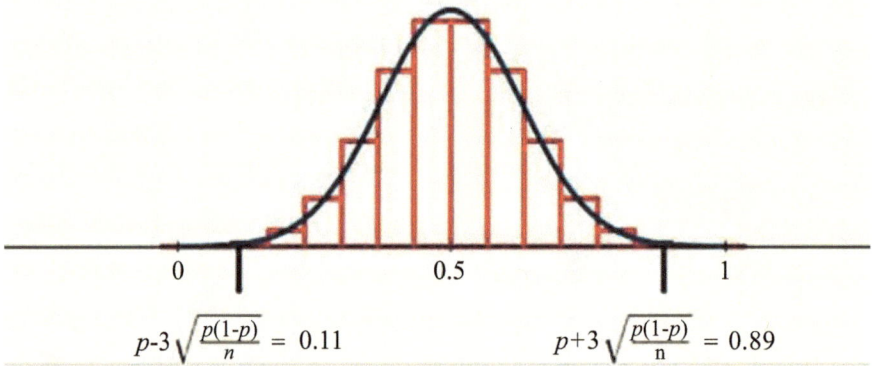

$p-3\sqrt{\frac{p(1-p)}{n}} = 0.11$ $p+3\sqrt{\frac{p(1-p)}{n}} = 0.89$

Fig. (11). Distribution of Sample Proportions for P [44].

Examples

Example 1

Assume that 38 percent of voters in a certain area support a particular bond issue. A total of 900 people will be polled to see whether they support the bond proposal.

a. Check that the sample percentage P^ calculated from 900 samples fits the criterion that the sampling distribution is nearly normal.
b. Calculate the likelihood that the sample proportion derived from a 900-person

sample will be within 5% of the genuine population proportion [44].

Solution

a. The information provided indicates that p=0.38, implying that q=1p=0.62. To begin, we calculate the mean and standard deviation (SD) of P^ using the following formulas [44]:

μP^ = p = 0.38 and σP^= pq/n = 0.38)(0.62)/900 = 0.01618

Then 3σP^ = 3 (0.01618) = 0.04854 $\approx$ 0.05 so

[p-3σP^, p +3σP^] = [0.38 - 0. 05, 0. 38 + 0.05] = [0. 33, 0. 43]

It is safe to infer that P^ is nearly normally distributed since it falls entirely inside the interval [0,1].

b. To be within 5% of the genuine population proportion of 0.38, you must be between [44]

0.38 - 0.05 = 0.33 and 0.38 = 0.05 = 0.43. Thus,

P(0.33 < P^ < 0.43) = P(0.33-μP^/ σP^< Z < 0.43-μP^/ σP^

= P(0.33-0.38 / 0.01618 < Z < 0.43-0.38 / 0.01618)

= P(-3.09 < Z < 3.09)

= P (3.09) - P(-3.09)

= 0.9990 - 0.0010 = 0.9980

Example 2

Let's say you know that 43% of Americans possess an iPhone. What is the likelihood that a random sample of 50 Americans owning an iPhone will be between 45 percent and 50 percent?

Solution

We know p=0.43 and n=50 for this issue. First, we should double-check our sampling distribution criteria for the sample percent [44].

np = 50(0.43) = 21.5 and n(1 − p) =50 (1-0.43) = 28.5 both are greater than 5.

Because all of the requirements are met, p^ sampling distribution will be essentially normal, with a mean of μ=0.43 and a standard deviation (SD) of [standard error] [44].

0.43(1-0.43)/50 ≈ 0.07.

$P(0.45 < p^\wedge < 0.5) = P\ (0.45\text{-}0.43/0.07 < p^\wedge - p\ /p(1\text{-}p)/n < 0.5 - 0.43\ /\ 0.07$

$= P(0.286 < Z < 1)$

$= P(Z < 1) - P(Z < 0.286)$

$= 0.8413 - 0.6126$

$= 0.2287$

As a result, if the genuine percentage of Americans who possess an iPhone is 43 percent, there is a 22.87 percent probability that when the sample size is 50, we will observe a sample proportion between 45 and 50 percent [44].

Example 3

According to one online shop, 90 percent of all purchases are sent within 12 hours after receipt. A customer group made 121 purchases of various sizes and times of day, with 102 sent within 12 hours.

a. Determine the percentage of products dispatched within 12 hours in the sample.
b. Confirm that the sample size is sufficient to presume that the sample percentage is normally distributed. Use p = 0.90 to represent the likelihood that the retailer's assertion is true.
c. Assuming the retailer's assertion is correct, calculate the likelihood that a sample size 121 would provide a sample percentage as low as the one found in this sample.
d. Conclude the retailer's assertion based on section (c) response.

Solution

a. The sample percentage is calculated by dividing the number of orders dispatched within 12 hours by the total number of orders in the sample [44]:

$p^\wedge = x\ /\ n = 102/121 = 0.84$

b. Since $p = 0.90$, $q = 1 - p = 0.10$, and $n = 121$,

σP^= (0.90)(0.10)/1210 = 0.027

Hence

[*p*-3σP^, *p* +3σP^] = [0.90 - 0.08, 0. 90 + 0.08] = [0.82, 0.98]

Because

[0.82, 0.98] ⊆ [0, 1].

To calculate probabilities associated with the sample percentage P^, the normal distribution is adequate.

c. we used the value of P^, and in part (b), we used the calculation (b) [44]:

$P(P^ \leq 0.84) = P(Z \leq 0.84 - \mu P^ / \sigma P^)$

$= P (Z \leq 0.84 - 0.90 / 0.027)$

$= P (Z \leq -2.20) = 0.0139$

When drawn from a population in which the real percentage is 0.90, a random sample of size 121 has only a 1.4 percent probability of yielding a sample proportion as the one that was seen, p^=0.84. [2] The calculation shows this, which demonstrates that the sample proportion observed is 0.90. The likelihood of this occurring is so remote that it is acceptable to assume that the true value of p is smaller than the reported 90 percent value.

CONCLUSION

To conclude, the mean and standard deviation of the sample mean is one of the most common and powerful tools used by researchers involved in data science and statistics. The Central Limit Theorem is a powerful tool for showing the distribution of data, outliers within the dataset and errors or potential outliers within a dataset based on the ends of the normal curve. Application is of this information to the sample proportion as well as to treat the samples within a small proportion is a wonderful way to show the character of a dataset as well as to convey information about the variance of the included data points.

CHAPTER 7

Estimation

Abstract: It is critically important for researchers to select the correct sample size to determine the population means. By understanding the difference between small and large sample sets, researchers can then construct intervals of confidence that assist in determining the population means. Confidence intervals are the margins of the error present within the dataset and are included to show the confidence of the researcher in the integrity of their dataset. Common confidence intervals are 90%, 95%, and 99%. The Z confidence level is calculated to show where the mean is likely to fall, and the T confidence level is used only when the sample size is smaller than 30 samples.

Keywords: Interval, Population mean, Sample estimation.

INTRODUCTION

By including sample sizes in their data analysis, researchers can simply describe the uncertainty associated with their dataset as well as to show the error margins present within the set. The distinctions between large and small sample sizes are typically set to encompass both the upper and lower ends of the dataset as well as the margins of error present within the data set. Confidence levels are also included with confidence intervals to show agreement with the assumptions of the dataset. Common confidence levels are 90%, 95%, and 99%. These levels are used to show the surety of the researchers in their measurements as well as to assist in predicting the characteristics of future datasets based on current understandings to ensure that the correct size sample is being used to determine the population means.

Construction of Confidence Intervals

The statistical inference technique refers to obtaining conclusions from data using statistical methods. As a result, the most crucial things to consider are testing hypotheses and concluding. As a branch of statistics, estimate theory is responsible for extracting parameters from data that have been contaminated by noise [46].

Calculating the values of parameters using measured and observed empirical data is a subfield of statistics and signal processing used to calculate the values of parameters in mathematical models. To measure and diagnose the real value of a function or a certain group of populations, the process of estimating must be carried out. It is carried out based on observations made on samples that provide a composite representation of the target population or function. When performing the estimate process, a variety of statistics are used [46].

In statistics, one of the most common applications is the estimation of population parameters using sample statistics. For example, a survey may be conducted to determine the percentage of adult citizens of a city who favor a proposal to construct a new sports stadium. A random sample of 200 persons was used to determine whether or not they endorsed the concept. As a result, 0.53 (106/200) of the persons in the sample agreed with the notion. The population percentage point estimate is defined as 0.53 (or 53 percent) divided by the total population. In this case, the estimate is a point estimate since it is composed of a single number or point [47].

It is very unusual for the actual population parameter to be the same sample value. When considering the hypothetical situation in which we surveyed the whole city's adult population, it is very implausible that precisely 53 percent of the population would support the notion. As an alternative, we may present a range of possible values for the parameter by using confidence intervals [47].

As a result, point estimates are often augmented with interval estimates or confidence ranges to provide a complete picture. Constructed by utilizing a technique that includes the population parameter for a set fraction of the time, confidence intervals include the population parameter. The pollster would arrive at the following 95 percent confidence interval, for example, if he or she utilized a procedure that included the parameter 95% of the time it was used. The pollster would thus conclude that anywhere between 46 percent and 60 percent of the public favors the initiative. In most cases, the media will publish this result by stating that 53 percent of the population supports the idea with a margin of error of 7 percent or less [47].

Interval Estimate *vs.* Point Estimate

To estimate population parameters, statisticians employ sample statistics. Sample means, for example, are used to get population means, whereas sample proportions are used to calculate population proportions [48].

A population parameter estimate may be stated in two ways:

1. An educated guess. A single value of a statistic is a point a parameter for a population is estimated. The sample mean x, for example, is a point estimate of

the population mean. The sample percentage p is a point estimate of the population proportion P, in the same way.

2. Estimated interval. A population parameter is between two values specified by an interval estimate. An interval estimate of the population mean, for example, is an x b. It denotes that the population mean exceeds a but falls short of b.

Intervals of Confidence

Statisticians use a confidence interval to represent the accuracy and uncertainty associated with a sampling process. There are three elements to a confidence interval.

- The interval itself.
- The confidence level.
- The parameter being estimated.

A sampling method's confidence level expresses the method's degree of uncertainty. The statistic and the margin of error together provide an interval estimate of the method's accuracy [48].

Consider the situation in which you are attempting to calculate an interval a parameter for a population is estimated. For the purpose of defining this interval estimate, a 95 percent confidence interval might be employed. It shows that, if we used the same sampling approach to pick other samples and produce different interval estimates, the true population parameter would fall inside the range indicated by the sample statistic + margin of error 95 percent of the time [48].

Because confidence intervals represent (a) the accuracy of the estimate and (b) the estimate's uncertainty, they are favored over point estimates [48].

Confidence Level

A confidence level is the probability portion of a confidence interval. The confidence level indicates how likely a sampling technique yields a confidence interval containing the genuine population parameter [49].

Learn how to interpret what a confidence level signifies in this article. Consider the following scenario: we collected all possible samples from a population and generated confidence intervals for each of them. The true population parameter would be included in certain confidence intervals but not in others, depending on the confidence interval [49]. It is implied that the genuine population parameter is present in 95 percent of the intervals if the confidence level is 95 percent; if the confidence level is 90 percent, it is implied that the population parameter is present in 90 percent of the intervals, and so on [49].

The Error Margin

It is the amount of time it takes for anything to go wrong. The margin of error is the range of values above and below the sample statistic in a confidence interval [49].

Consider the case when a local newspaper conducts an election poll and finds that the independent candidate will obtain 30% of the vote. According to the publication, the poll had a margin of error of 5% and a confidence level of 95%. As a consequence of these results, the following confidence interval has been calculated: We have a 95% confidence level that the independent candidate will earn between 25% and 35% of the vote [49].

Interval estimates, but not confidence intervals, are reported in many public opinion polls. They give you the margin of error but not the degree of confidence. You must be familiar with both to properly understand survey data. If the confidence level is high (say, 95 percent), we are far more inclined to believe survey results than low (say, 50 percent) [49].

$$\bar{x} = \frac{1}{N} \sum_{i=1}^{N} x_i.$$

Estimator

The sample mean is an estimator for the population mean [50].

An estimator is a statistic that estimates a population statistic. An estimator may also be thought of as the rule that generates an estimate. For example, the sample mean(x) estimates the population mean [50].

The estimator is the amount that is being estimated (*i.e.*, the one you wish to know). For example, suppose you needed to discover the average height of pupils at a 1000-student school. You measure a group of 30 youngsters and discover that the average height is 56 inches. It is the estimate for your sample mean. Using the sample mean, you estimate the population mean (your estimator) to be about 56 inches [50].

Interval vs. Point Estimator

A range of values (such as a confidence interval) or a single number may be used as estimators (like the standard deviation) (SD). An interval estimate is when an estimator is a range of values. You might add a confidence interval of a few inches each way to the height example above, say 54 to 58 inches. A point estimate is when a single figure such as 56 inches, is used [50].

Types of Estimators

Estimators may be characterized in a variety of ways:

A statistic that is either an exaggeration or an underestimate is biased.

Efficient: a statistic with modest variances (the "best" statistic has the least possible variance). Inefficient estimators may also provide decent results, but they often need considerably larger sample sizes.

Invariant statistics are resistant to transformations, such as simple data shifts.

Shrinkage is used to describe a raw estimate enhanced by merging it with additional data. The James-Stein estimator is another option.

Sufficient: a statistic that estimates the population parameter as if all the data from all feasible samples were known.

Unbiased: a statistic that neither underestimates nor overestimates the situation [50].

WHAT IS STANDARD ERROR (SDE)?

The standard error is a calculation of a statistic's standard deviation (SD). The standard error (SE) is significant because it calculates other statistics such as confidence intervals and margins of error [51]. Fig. (1) shows the population parameter and sample statistics.

Standard Deviation (SD) of Sample Estimates

To estimate population parameters, statisticians employ sample statistics. Of course, the value of a statistic will vary from one sample to the next.

The standard deviation (SD) of a statistic is a measure of its variability. The formulae for determining the standard deviation (SD) of statistics from simple random samples are shown in the table below. These calculations are valid when the population size is substantially bigger (at least 20 times larger) than the sample size [51]. Fig. (**2**) shows the statistics and their standard deviation.

Population parameter	Sample statistic
N: Number of observations in the population	n: Number of observations in the sample
N_i: Number of observations in population i	n_i: Number of observations in sample i
P: Proportion of successes in population	p: Proportion of successes in sample
P_i: Proportion of successes in population i	p_i: Proportion of successes in sample i
μ: Population mean	$\bar{x}$: Sample estimate of population mean
$μ_i$: Mean of population i	$\bar{x_i}$: Sample estimate of $μ_i$
σ: Population standard deviation	s: Sample estimate of σ
$σ_p$: Standard deviation of p	SE_p: Standard error of p
$σ_{\bar{x}}$: Standard deviation of $\bar{x}$	$SE_{\bar{x}}$: Standard error of $\bar{x}$

Fig. (1). Population parameter and sample statistics [51].

Statistic	Standard Deviation
Sample mean, $\bar{x}$	$σ_{\bar{x}} = σ / \sqrt{n}$
Sample proportion, p	$σ_p = \sqrt{P(1 - P) / n}$
Difference between means, $\bar{x}_1 - \bar{x}_2$	$σ_{\bar{x}_1 - \bar{x}_2} = \sqrt{σ^2_1 / n_1 + σ^2_2 / n_2}$
Difference between proportions, $\bar{p}_1 - \bar{p}_2$	$σ_{p_1 - p_2} = \sqrt{P_1(1-P_1) / n_1 + P_2(1-P_2) / n_2}$

Fig. (2). Statistics and their standard deviation [51].

Note: You must know the value of one or more population parameters to calculate the standard deviation (SD) of a sample statistic. To calculate the standard deviation (SD) of the sample mean ($σ_{\bar{x}}$), for example, you'll need to know the population variance (σ) [51].

Standard Error (SE) of Sample Estimates

Unfortunately, the values of population parameters are often unknown, making computing the standard deviation (SD) of a statistic unfeasible. Use the usual error if this happens.

The standard error is calculated using sample data that are known. The table below explains how to calculate the standard error for simple random samples with a population size of at least 20 times the sample size [51]. Fig. (**3**) shows the statistics and their standard error.

Statistic	Standard Error
Sample mean, $\bar{x}$	$SE_{\bar{x}} = s / \sqrt{n}$
Sample proportion, p	$SE_p = \sqrt{[\, p(1-p)/n \,]}$
Difference between means, $\bar{x}_1 - \bar{x}_2$	$SE_{\bar{x}_1 - \bar{x}_2} = \sqrt{[\, s^2_1/n_1 + s^2_2/n_2 \,]}$
Difference between proportions, $\bar{p}_1 - \bar{p}_2$	$SE_{p_1 - p_2} = \sqrt{[\, p_1(1-p_1)/n_1 + p_2(1-p_2)/n_2 \,]}$

Fig. (3). Statistics and their standard error [51].

The standard error (SE) equations are similar to the standard deviation (SD) equations, with one exception: the standard error equations employ statistics, while the standard deviation (SD) equations use parameters. In the standard error equations, p is used instead of P, and s is used instead of σ [51].

Margin of Error

The margin of error is the range of values above and below the sample statistic in a confidence interval. Let's say we wanted to know what proportion of folks workout every day. We could create a sample design that ensures our sample estimate doesn't deviate from the genuine population value by more than 5% (the margin of error) 90% of the time (the confidence level) [52].

How to Calculate the Error Margin

Either of the following formulae may be used to calculate the margin of error [52].

The error margin equals the critical value multiplied by the statistic's standard deviation (SD) [52].

The error margin equals the critical value multiplied by the statistic's standard error [52].

Use the first equation to calculate the margin of error if you know the statistic's standard deviation (SD). Use the second equation if the first doesn't work. We have covered how to calculate standard deviation (SD) and standard error [52].

What is the Critical Value and How Do I Find it?

A factor used to calculate the margin of error is called the crucial value. When the statistic's sample distribution is normal or nearly normal, this section explains how to determine the critical value.

The crucial value may be represented as a t score or a z-score when the sample distribution is almost normal. Follow these procedures to determine the crucial value.

Calculate alpha (α) using the formula: $\alpha = 1 - (\text{confidence level} / 100)$.

Calculate the critical probability (p*) as follows: $p* = 1 - \alpha/2$

Find the z-score with a cumulative probability equal to the critical probability (p*) to represent the crucial value as a z-score.

Follow these methods to represent the crucial value as a t statistic.

Calculate the degrees of freedom (DF). DF equals the sample size minus one when calculating a mean score or a percentage from a single sample. The degrees of freedom may be computed differently in various applications. We'll go through those calculations as they arise.

The crucial t statistic (t*) is a t statistic with degrees of freedom of DF and a cumulative probability of p* [52].

The T-Score and the Z-Score are two different types of scores.

Should you use a t statistic or a z-score to indicate the important value? The population standard deviation (SD) is one technique to solve this issue.

- Use the z-score if the population standard deviation (SD) is known.
- Use the t statistic if the population standard deviation(SD) is unknown.

Another strategy is to concentrate on sample size.

- Use the z-score if the sample size is big. (The central limit theorem may be used to determine whether or not a sample is "big.")
- Use the t statistic if the sample size is small.

In practice, researchers use a combination of the criteria mentioned above. When the population standard deviation (SD) is known, and the sample size is high, we utilize z-scores on this site. Unless the sample size is tiny and the underlying distribution is not normal, we employ t statistics [52].

We cannot be certain that the statistic's sampling distribution will be normal if the sample size is small and the population distribution is not normal. The t statistic and the z-score should not be utilized to derive critical values in this case [52].

The crucial z-score may be found using the Normal Distribution Calculator, while the critical t statistic can be found using the t Distribution Calculator. A graphing calculator or normal statistical tables may also be used (found in most introductory statistics texts) [52].

What is a Confidence Interval, and How Does It Work?

To characterize the level of uncertainty associated with a sample a parameter for a population is estimated, statisticians use the term "confidence interval" [49].

Confidence Intervals and How to Interpret Them

Assume that the population mean is more than 100 and less than 200, according to a 90% confidence interval. What do you think of this statement?

Some individuals interpret this to suggest that the population mean will most likely fall between 100 and 200. This isn't true. Like every other population characteristic, the population means is a constant, not a random variable. It stays the same. A constant's chance of falling inside a particular range is always 0.00 or 1.00 [49].

A sampling method's confidence level defines the amount of uncertainty associated with it. Assume we selected several samples using the same sampling approach and computed a different interval estimate for each sample. The genuine population parameter would be included in certain interval estimations but not others. A 90% confidence level indicates that we may anticipate the population parameter to appear in 90% of the interval estimates; a 95% confidence level means that the parameter will appear in 95% of the intervals; and so on [49].

Data Requirements for Confidence Interval

You need three pieces of information to express a confidence interval.

- Statistic
- Margin of Error
- Confidence Level

The sample statistic plus the margin of error establishes the confidence interval range given these inputs. The confidence level defines the amount of uncertainty associated with the confidence interval.

The margin of error is often not stated and must be calculated. The method for calculating the margin of error was previously outlined [49].

What is a Confidence Interval, and How Do I Make One?

A confidence interval is constructed in four stages.

Determine a representative statistic. Choose the statistic you'll use to estimate a population parameter (*e.g.,* the sample mean, sample percentage).

Choose a degree of assurance. As discussed in the previous section, the confidence level represents the uncertainty of a sampling process. Researchers often employ confidence levels of 90 percent, 95 percent, or 99 percent, although any proportion may be utilized [49].

Calculate the error margin. The margin of error may be specified while working on a homework assignment or a test question. However, you'll often need to calculate the margin of error using one of the formulae below.

The margin of error is equal to the critical value multiplied by the standard deviation (SD) of the statistic.

The margin of error is equal to the critical value multiplied by the standard error of the statistic.

See how to calculate the margin of error for further information.

The confidence interval must be specified. The confidence level expresses the amount of uncertainty. The following equation specifies the confidence interval's range [49].

Sample statistic + margin of error = confidence interval

In the following section, you'll see an example issue that uses the four stages above to create a 95 percent confidence interval for a mean score. The next classes go through this subject in further depth [49].

Bias and Error

Allow for the possibility that 0 (the Greek letter "theta") is the value of the population parameter we are interested in. In statistics, we would denote the estimate as a value of θ. (read theta-hat). We are aware that the estimated population parameter will seldom be equal to the real population parameter θ. There is a certain amount of mistake connected with this. This kind of mistake is denoted by the letter $e(x) = \theta^\wedge (x) - \theta$ [47].

Every measurement has a certain amount of inaccuracy connected with it. Random mistakes may occur in any data collection and are referred to as non-systematic errors in certain circles [47]. Random mistakes may occur due to inaccurate estimates of data values, imprecision of equipment, and other factors. Consider the following example: if you measure lengths using a ruler, random mistakes will occur in each measurement due to judging where the length falls between two lines. Bias is sometimes referred to as a systematic mistake in certain circles. It is biased when a value is continuously underestimated or overstated in data collection. Another source of bias is the failure to consider a correction factor or the use of instruments that are not correctly calibrated, among other things. Due to bias, a sample mean that is either lower or higher than the genuine mean is produced (Fig. **4**) [47].

MSE stands for Mean Squared Error

The mean squared error (MSE) of θ is the anticipated value of the squared errors multiplied by the number of observations. On average, it highlights how far away the collection of estimates is from the single parameter under consideration [47].

Assume that the parameter is the bulls-eye of a target. The estimator is the process of firing arrows at the target. The individual arrows represent estimations of the parameter (samples). MSE indicates the average distance between the arrows and the bull's-eye in this instance, whereas low MSE indicates the average distance between the arrows and the bull's-eye. The arrows may be grouped or not. If, for example, all of the arrows land on the same spot but miss the target by a wide margin, the MSE is still a significant amount. But when the MSE is relatively low, the arrows are likely to be strongly crowded, as seen in the example above (then highly dispersed) [47].

Effects of sample bias coefficient ρ

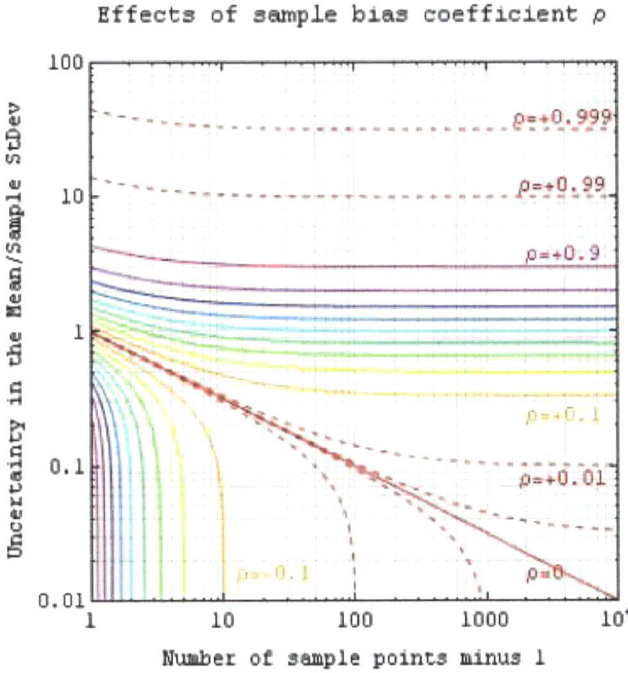

Sample Bias Coefficient: An estimate of expected error in the sample mean of variable A, sampled at N locations in a parameter space x, can be expressed in terms of sample bias coefficient ρ — defined as the average auto-correlation coefficient over all sample point pairs. This generalized error in the mean is the square root of the sample variance (treated as a population) times $\frac{1+(N-1)\rho}{(N-1)(1-\rho)}$. The $\rho = 0$ line is the more familiar standard error in the mean for samples that are uncorrelated.

Fig. (4). Effects of sample bias coefficient p [47].

A viable technique for estimating the mean of a population for which a census is impossible, such as the average height of all 18-year-old males in the nation, is to take a sample, calculate the mean $\bar{x}$, an nd estimate the unknown number μ by the known number $\bar{x}$. If the average height of 100 randomly chosen males aged 18 is 70.6 inches, we may conclude that the average height of all 18-year-old men is (at least roughly) 70.6 inches [47].

Point estimation is the process of estimating a population parameter using a single number; in this situation, the statistic $\bar{x}$ is a point estimate of the parameter μ.

A single number corresponds to a single point on the number line, hence the phrase [47].

A point estimate has the disadvantage of not indicating the estimate's reliability. In this chapter, on the other hand, we will learn about interval estimation. In short, while calculating a population mean, we utilize a formula to calculate a number E, also known as the margin of error of the estimate, from the data and construct the interval $[x^- E, x^- +E]$. We do this in such a manner that the unknown parameter μ appears in a specific percentage of all intervals built from sample data using this technique, say 95%. A 95 percent confidence interval is one example of such an interval μ [47].

Continuing with the average height of 18-year-old males as an example, imagine the sample of 100 men with x^-=70.6 inches also had a sample standard deviation (SD) of s=1.7 inches. As a result, E = 0.33, and we can say with 95 percent certainty that the average height of all 18-year-old males is within the range of 70.6 ±0.33 inches, *i.e.*, the average is between 70.27 and 70.93 inches. If the sample data were from a smaller sample, say 50 males, the reduced dependability would be reflected in a larger 95 percent confidence interval, making the estimate less exact. For the same sample statistics with n = 50, the 95 percent confidence interval is 70.6 ±0.47 inches or 70.13 to 71.07 inches in this case [47].

Sample Size and Estimates

As a result, we demonstrate how to compute the smallest possible sample size required to estimate a population mean μ and population proportion p.

Determining the Sample Size is Necessary to Estimate the Population Mean

A sample must be obtained before a point estimate can be calculated and a confidence interval constructed. In many cases, the number of data values required in a sample to achieve a certain degree of confidence within a given error must be decided before the sample is taken [47]. The results may be useless if the sample is too small, and the sample will be too large, wasting both time and money in the sampling process. Given the desired confidence level and an observed standard deviation (SD), the following text illustrates how to calculate the smallest sample size required to generate an estimate [47].

The margin of error, denoted by the letter E, is the largest possible gap between a point estimate and the value of the parameter it is attempting to estimate. To compute E, we must first determine the required confidence level (Z α/2) and the population standard deviation (SD) σ. When the sample standard deviation (SD)

(s) is less thann>30, the sample standard deviation (SD) (s) can be used to approximate the population standard deviation (σ).

$$E = Z \, \alpha/2 \, \frac{\sigma}{\sqrt{n}}$$

Two variables in the formula could be changed to alter the size of the error (E): the level of confidence (Z α/2) and the sample size (n). The standard deviation (SD) (σ) is a given and cannot be changed under any circumstances [47].

The margin of error (E) grows in proportion to the increase in confidence. The confidence level would have to be reduced to guarantee that the margin of error is as little as possible. Hence, lowering the confidence to decrease the mistake is not viable [47].

The margin of error shrinks in direct proportion to the increase in sample size (n). The issue now is: how much of a sample is required for a certain mistake to be detected? Begin by calculating the equation for E in terms of n, which may be found in the table below (Fig. **5**):

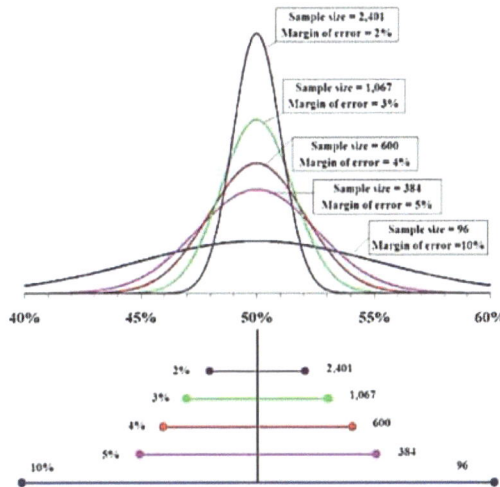

Sample size compared to margin of error: The top portion of this graphic depicts probability densities that show the relative likelihood that the "true" percentage is in a particular area given a reported percentage of 50%. The bottom portion shows the 95% confidence intervals (horizontal line segments), the corresponding margins of error (on the left), and sample sizes (on the right). In other words, for each sample size, one is 95% confident that the "true" percentage is in the region indicated by the corresponding segment. The larger the sample is, the smaller the margin of error is.

Fig. (5). Margin of Error [47].

$$n = (Z \, \alpha/2 \, \sigma / E)^2$$

where $Z \, \alpha/2$ represents the crucial z score depending on the desired confidence level, E represents the required margin of error and represents the population standard deviation (SD).

Because the standard deviation (SD) of the population is often unknown, the sample standard deviation (SD) from a prior sample of size $n \geq 30$ may be used as an approximation to s in certain situations. We can now solve for n to get the sample size that would be most suited for achieving our objectives. It is important to note that the result obtained by using the formula for sample size is seldom a full integer. The sample size must be an integer, thus rounding up to the nearest bigger whole number wherever possible [47].

Examples

Example

Take, for example, a statistics final in which the results are normally distributed with a standard deviation (SD) of 10 points. Create a 95 percent confidence interval with an inaccuracy of no more than 2 percentage points in your data.

Solution [47]

$$Z_{0.025} = 1.645$$
$$E = 2$$
$$\sigma = 10$$
$$n = (1.645(10) / 2)^2 = 8.225^2 = 67.75$$

As a result, a sample of 68 people must be collected to provide a 95 percent confidence interval with an error of no more than 2 percentage points [47].

Large Sample Estimation of a Population Mean

The Central Limit Theorem states that, when seen as a random variable, the sample mean $\bar{X}$ is normally distributed with mean $\mu \bar{X} = \mu$ and standard deviation (SD) $\sigma \bar{X} = \sigma/n$ for large samples (samples of size n 30). According to the Empirical Rule, we need to travel two standard deviations (SD) from the mean to collect 95% of the $\bar{X}$ values provided by sample following sample. 1.960 standard deviations, (SD) or $E = 1.960 \, \sigma/n$, is a more accurate distance depending on $\bar{X}$ normalcy [53].

Because 95 percent of the values of X⁻ fall in the range [μ-E, μ+E], if we add a "wing" of length E to either side of the point estimate x⁻ 95 percent of the intervals produced by the winged dots μ include. As a result, the 95 percent confidence interval is x⁻ ±1.960 σ/n. Fig. (**6**), 1.960 will alter depending on the amount of confidence, such as 90 percent or 99 percent, but the concept remains the same [53].

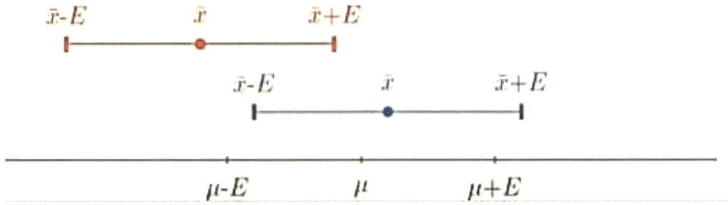

Fig. (6). Winged dots capture the population mean [53].

The intervals created by a computer simulation of taking 40 samples from a normally distributed population and computing the 95 percent confidence interval for each one are shown in Fig. (**7**) "Computer Simulation of 40 95 percent Confidence Intervals for a Mean." We predict roughly (0.05)(40)=2 of the intervals generated this way to fail to contain the population mean μ, and two of the intervals depicted in red do in this simulation [53].

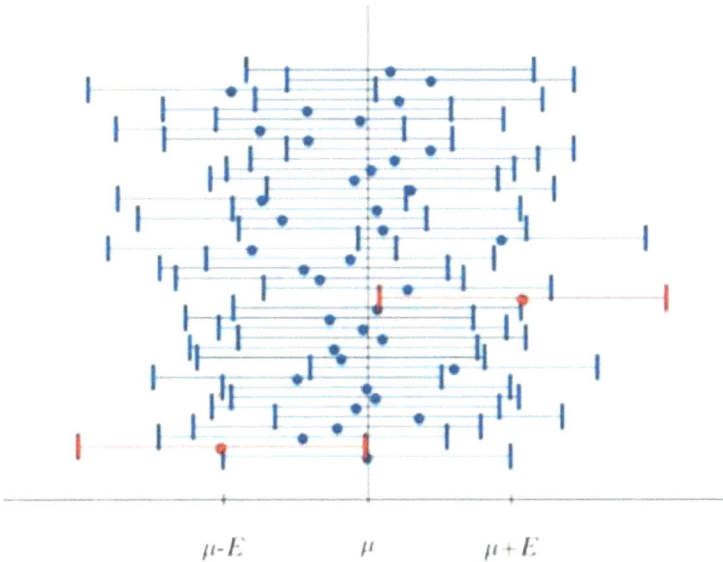

Fig. (7). Computer Simulation of 40 95 percent Confidence Intervals for a Mean [53].

When the center region indicated by the level of confidence is removed, it is normal to practice calculating the level of confidence in terms of the area in the two tails of the X^- Distribution [53]. It is shown in Fig. (**8**), which depicts the overall condition, and (Fig. **9**), which depicts 95 percent certainty. Remember that the z-value that cuts off a right tail of area c is indicated z_c from previous section "Tails of the Standard Normal Distribution" in Chapter 5, "Continuous Random Variables." As a result, in the example, the number 1.960 is:

$Z_{0.025}$, which is $Z\alpha/2$ for $\alpha = 1 - 0.95 = 0.05$.

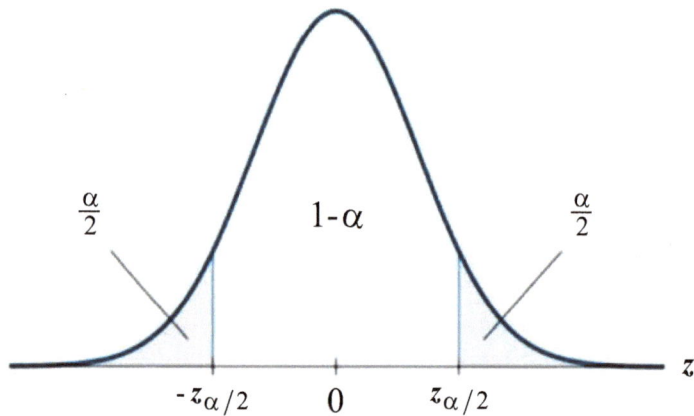

For 100 $(1 - \alpha)$% confidence the area in each tail is $\alpha / 2$.

Fig. (8). Tails of distribution [53].

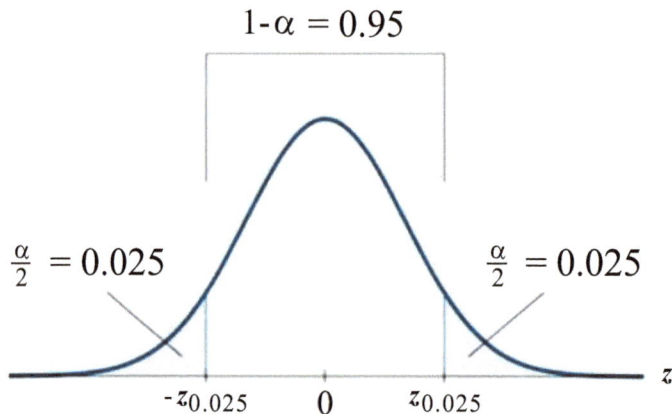

For 95% confidence the area in each tail is $\alpha / 2 = 0.025$.

Fig. (9). Certainty depiction [53].

The degree of confidence may range from 0 to 100 percent, although the most typical numbers are probably 90% (α=0.10), 95% (α=0.05), and 99 percent (α=0.01).

In general, E=z $\alpha/_2(\sigma/n)$ for a 100(1-α) percent confidence interval, hence the confidence interval formula is x¯$\pm$z $\alpha_{/2}(\sigma/n)$. While the population standard deviation (SD) is occasionally known, it is usually not. If not, for n $\geq$ 30, it is typically safe σ to use the sample standard deviation (SD) s as an approximation.

Large Sample 100 (1 - α) % Confidence Interval for a Population Mean

If σ is known: $\bar{x} \pm z \, \alpha_{/2} \left(\frac{\sigma}{\sqrt{n}}\right)$

If σ is unknown: $\bar{x} \pm z \, \alpha_{/2} \left(\frac{s}{\sqrt{n}}\right)$

A sample is considered large when n $\geq$ 30.

As mentioned earlier, the number $E = z \, \alpha_{/2} \, \sigma \, \sqrt{n}$ or $E = z \, \alpha_{/2} \, s \, \sqrt{n}$ is called the *margin of error* of the estimate.

Example

The sample mean is 35, while the sample standard deviation (SD) is 14. Creating a 98 percent confidence interval for the population mean using these data. Interpret what it means.

Solution

For confidence level 98%, α = 1 - 0.98 = 0.02, so $z\alpha_{/2}$ = $z_{0.01}$. From Fig. (**13**) of chapter 7 "Critical Values of t distribution",

we read directly that $z_{0.0.1}$ = 2.326. Thus [53],

$$\bar{x} \pm z \, \alpha_{/2} \left(\frac{s}{\sqrt{n}}\right) = 35 \pm 2.326 \, (14/\sqrt{49}) = 35 \pm 4.652 \approx 35 \pm 4.7$$

We have 98 percent confidence that the population mean is in the range [30.3, 39.7], in the sense that 98 percent of all intervals created from the sample data in this way would include μ [53].

Small Sample Estimation of a Population Mean

The preceding section's confidence interval calculations are based on the Central Limit Theorem, which states that $\bar{X}$ is normally distributed with mean and standard deviation (SD)σ/n for large samples [8]. The Central Limit Theorem does not apply when the population mean μ is computed using a small sample (n <30). To continue, we must first assume that the numerical population from which the sample is drawn has a normal distribution [8]. If this requirement is met, the previous formula $\bar{x} \pm z\alpha/2(\sigma/n)$ may still be used to generate a 100(1-α) percent confidence interval for when the population standard deviation (SD) is known [53].

When the population standard deviation (SD) is unknown, and the sample size n is small, the normal approximation is no longer applicable when the standard deviation (SD) s is substituted [53]. An alternative distribution, known as the Student's t-distribution with n-1 degrees of freedom, is used as a solution. The Student's t-distribution is similar to the standard normal distribution in that it is centered at 0 and has the same qualitative bell shape. Still, it has heavier tails, as shown in Fig. (10) Students t-distribution, where the t-distribution with two degrees of freedom meets the dashed vertical line at the lowest point, the next curve (in blue) is the t-distribution with five degrees of freedom [53]. The third curve (in brown) is the Student's t-distribution closely mimics the usual normal distribution as the sample size n rises, as seen in the image. Although each number of n has a distinct t-distribution, if the sample size reaches 30 or more, it is usually appropriate to use the ordinary, normal distribution instead, which we shall do throughout this article.

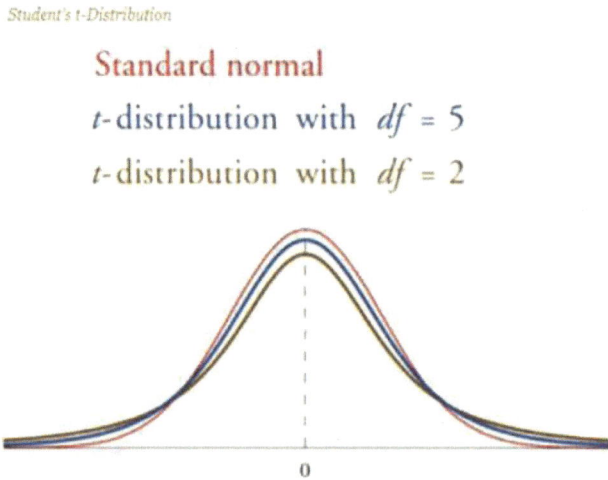

Fig. (10). Students t-distribution [53].

The sign t_c denotes the value that cuts off a right tail of area[1] c in the standard normal distribution, just as the symbol z_c denotes the value in the standard normal distribution. The following formulae f[2]or confidence intervals result from this [53].

Small Sample 100 (1 -α) % Confidence Interval for a Population Mean [53]

If σ is known: $\bar{x} \pm z\alpha_{/2}$ (σn)

If σ is unknown: $\bar{x} \pm t\alpha_{/2}$ (sn) (degrees of freedom $df = n - 1$)

The population must be normally distributed.

A sample is considered small when n < 30.

Example 1

A sample size of 15 selected from a normally distributed population has a sample mean of 35 and a standard deviation (SD) of 14. Construct and analyze a 95 percent confidence interval for the population mean.

Solution:

Because the population is normally distributed, the sample is tiny, and the standard deviation (SD) of the population is unknown, the formula to use is:

$$\bar{x} \pm t\,\alpha_{/2}\left(\frac{s}{\sqrt{n}}\right)$$

Confidence level 95% means that α = 1 - 0.95 = 0.05 so $\alpha_{/2}$ = 0.025. Since the sample size is n = 15, there are n - 1 = 14 degrees of freedom, By Fig. (**13**) of chapter 7 "Critical Values of t distribution" $t_{0.025}$ = 2.145. Thus

$$\bar{x} \pm t\,\alpha_{/2}\left(\frac{s}{\sqrt{n}}\right) = 35 \pm 2.145\,(14\,/\sqrt{15}) = 35 \pm 7.8$$

One may be 95% confident that the true value of μ contained in the interval (35 - 7.8, 35 + 7.8) = (27. 2, 42.8) [53].

Example 2

The mean GPA of 12 students from a big institution is 2.71, with a sample standard deviation (SD) of 0.51. Construct a 90% confidence interval for the average GPA of all university students. Assume a normal distribution for the numerical population of GPAs from which the sample is drawn.

Solution

Because the population is normally distributed, the sample is tiny, and the standard deviation (SD) of the population is unknown, the formula to use is:

$$\overline{x} \pm t\,\alpha_{/2}\left(\frac{s}{\sqrt{n}}\right)$$

Confidence level 90% means that $\alpha = 1 - 0.90 = 0.10$ so $\alpha_{/2} = 0.05$. Since the sample size is $n = 12$, there are $n - 1 = 11$ degrees of freedom, By Fig. (**13**) of chapter 7 "Critical Values of t distribution" $t_{0.05} = 1.796$. Thus,

$$\overline{x} \pm t\,\alpha_{/2}\left(\frac{s}{\sqrt{n}}.\right) = 2.71 \pm 1.796\,(0.51\,/\sqrt{12}) = 2.71 \pm 0.26$$

One may be 90% confident that the true average GPA of all the students at the university is contained in the interval $(2.71 - 0.26, 2.71 + 0.26) = (2.45, 2.97)$ [53].

Determining Sample Size Required to Estimate Population Proportion (*p*)

For choosing sample size to estimate a percentage (p), the calculations are identical to those estimating the mean (μ). The margin of error, E, in this example is calculated using the following formula [8]:

$$E = Z\,\alpha_{/2}\sqrt{\frac{p'q'}{\sqrt{n}}})$$

Where

- $p' = x/n$ is the point estimate for the population proportion
- x is the number of successes in the sample
- n is the number in the sample; and
- $q' = 1 - p'$

After that, we may solve for the smallest sample size n required to estimate p as follows [53]:

$$n = p'q'\,(Z\,\alpha_{/2}\,/\,E)^2$$

Example

According to the Mesa College mathematics department, several students are placed in non-transfer level courses and only need a 6-week refresher course rather than a full semester-long course. Suppose it is estimated that around 10% of students fall into this group. How many students does the department need to poll to be 95 percent certain that the genuine population percentage is within ±5% of the estimated proportion?

Solution [53]

Z = 1.96

E = 0.05

p' =0.1

q' = 0.9

n = (0.1) (0.9) (1.96 / 0.05)² ≈ 138.3

As a result, a sample of 139 people must be collected to generate a 95 percent confidence interval with less than ±5% error.

Estimating the Target Parameter: Point Estimation

Point estimation is the process of calculating a single value from a sample of data that serves as the "best estimate" of an unknown population parameter using sample data [47].

A sample of data can be used to "estimate" or "guess" information about data from a larger population, which is known as inferential statistics. Point estimation is the process of calculating a single value or point (also known as a statistic) from a sample of data to get the "best estimate" of an unknown population parameter from the data [47]. An a parameter for a population is estimated, such as the mean, that is given as a single number is known as a point estimate of the mean. It is the sample mean (x⁻) that provides the most accurate point estimate of a population mean (μ) (Fig. **11**) [47].

Using a simple random sampling of a population, we can estimate or guess information about the data from a population by using point estimators, such as the sample mean. This graphic representation depicts the process of picking members of a bigger group of individuals to represent that larger group based on a random number issued to them [47].

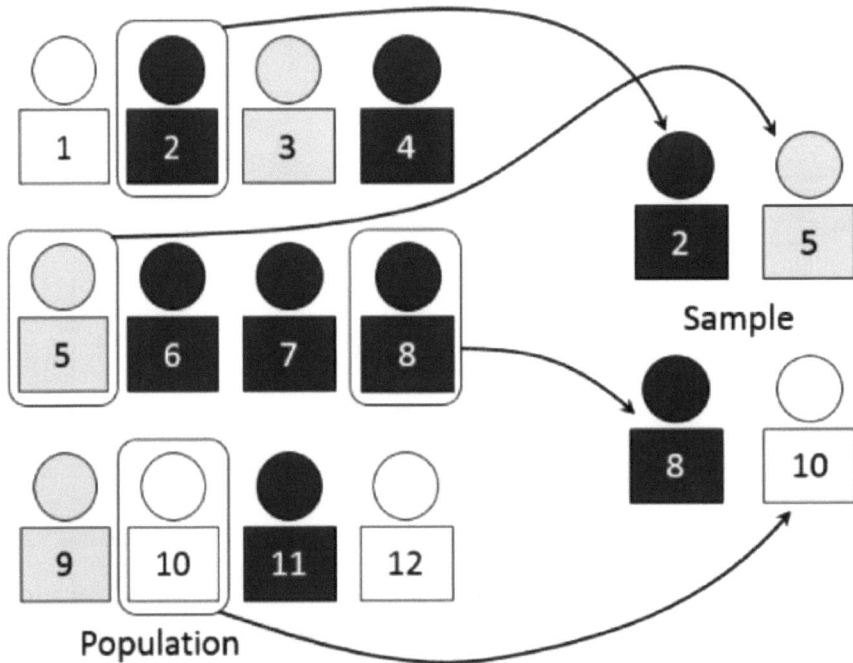

Fig. (11). Point estimation [47].

Maximum Likelihood

The maximum-likelihood estimation technique is one of the most often used methods of estimating the parameters of a statistical model (MLE). Maximum-likelihood estimation is a statistical technique that provides estimates for the model's parameters when applied to a data set and provide a statistical model. The maximum likelihood technique is similar to several well-known estimate methods in statistics. Consider the following scenario: a researcher is interested in adult female penguins but cannot measure the heights of every single penguin in a colony due to financial or time restrictions. Accordingly, assuming that the heights are normally (Gaussian) distributed with a previously unknown mean and variance, the mean and variance may be calculated using MLE using just a sample of the whole population's heights. To do this, the mean and variance would be taken into consideration [47].

In general, given a fixed set of data and an underlying statistical model, the maximum likelihood technique picks the set of model parameters that maximizes the likelihood function, which is the likelihood function of the data. A unified approach to estimate is provided by maximum-likelihood estimation, well-defined in the case of the normal distribution and various additional issues. On the other hand, maximum-likelihood estimators are inappropriate or do not exist in certain intricate scenarios [47].

Linear Least Squares (LLS)

The linear least-squares method is yet another common estimating technique to use. Linear least squares is a statistical modeling technique used to fit a statistical model to data in situations where the intended value produced by the model for every data point can be stated linearly in terms of the model's unknown parameters (as in regression). The fitted model that is produced may summarize the data, estimate unobserved values from the same system, and understand the processes at the root of the problem [47].

Linear least squares is a mathematical problem that involves approximately solving an over-determined system of linear equations, where the best approximation is defined as the one that minimizes the sum of squared differences between the data values and their corresponding modeled values [47]. Linear least squares is a mathematical problem that involves approximately solving an over-determined system of linear equations. The methodology is called "linear" least squares since the assumed function is linear in the parameters that must be evaluated when using this method. As a kind of regression analysis, linear least squares issues relate to a statistical model known as linear regression, which develops as a specific form of linear regression analysis. An ordinary least squares model is one of the most fundamental types of such a model [47].

Estimating the Target Parameter: Interval Estimation

Interval estimate is the process of calculating an interval of potential (or likely) values for an unknown population parameter by analyzing a sample of data.

Interval estimate is the process of calculating an interval of potential (or likely) values for an unknown population parameter by analyzing a sample of data. The most often used interval estimate methods are confidence intervals (based on the frequents technique) and credible intervals (a Bayesian method) [47].

Other ways to interval estimate that are often used include as follows:

- Tolerance intervals
- Prediction intervals are mostly utilized in Regression Analysis and other fields.
- Intervals of likelihood

Example

Why is it difficult to establish an appropriate confidence interval for an unknown population mean μ when we don't know the population standard deviation σ is? To do this, we must make educated guesses based on the facts. In addition, we must check three criteria related to the data:

1. A simple random sample of the size n from the population of interest was used to gather the information.

2. Normal distribution is used to describe the data from the population, and it includes the mean and standard deviation. These are two unknown parameters.

3. The approach for establishing a confidence interval is based on the assumption that each observation is independent of the others.

It is assumed that the sample mean $\bar{x}$ is distributed normally, with a mean and standard deviation σn. Because we don't know σ what is, we make an educated guess based on the sample standard deviation s. As a result, we may estimate the standard deviation of $\bar{x}$ by utilizing the standard error of the sample mean, which is denoted by sn [47].

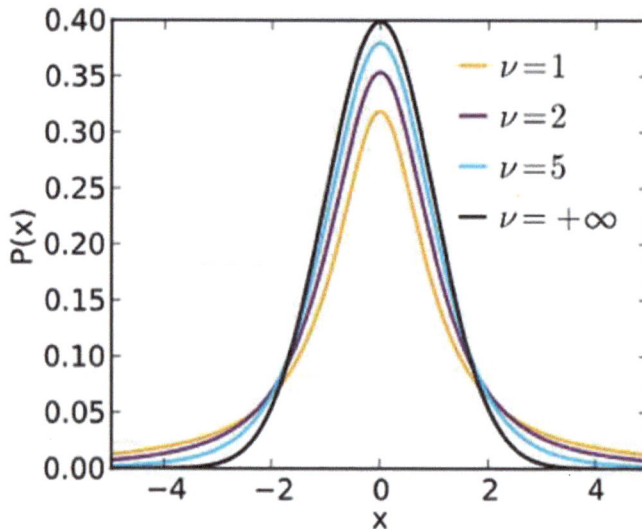

t-Distribution: A plot of the t-distribution for several different degrees of freedom.

Fig. (12). t-distribution [47].

The *t* Distribution

When we don't know what to say with σn. we utilize sn. The resultant statistic, t, has a distribution that is not Normal and is consistent with the t-distribution. [2] Each sample size n has a unique t-distribution, seen in the (Fig. **12**). To define a particular t-distribution, we utilize its degrees of freedom, indicated by df and df = n-1 [47].

If we wanted to estimate the population mean, we could now put everything we've learned together in one place [2]. First, choose a basic random sample from a population with an unknown mean to make conclusions. A confidence interval may be determined using the following formula:

$$x \pm t * \frac{s}{\sqrt{n}}$$

Where t* is the critical value for the t(n-1) distribution (Fig. **13**) [47].

df	0.2	0.1	0.05	α (Two-Tailed) 0.02	0.01	0.001	0.0001
1	3.078	6.314	12.706	31.821	63.657	636.619	6366.198
2	1.886	2.920	4.303	6.695	9.925	31.598	99.992
3	1.638	2.353	3.182	4.541	5.841	12.924	28.000
4	1.533	2.132	2.776	3.747	4.604	8.610	15.544
5	1.476	2.015	2.571	3.365	4.032	6.869	11.178
6	1.44	1.943	2.447	3.143	3.707	5.959	9.082
7	1.415	1.895	2.365	2.998	3.499	5.408	7.885
8	1.397	1.860	2.306	2.896	3.355	5.041	7.120
9	1.383	1.833	2.262	2.821	3.250	4.781	6.594
10	1.372	1.812	2.228	2.764	3.169	4.587	6.211
11	1.363	1.796	2.201	2.718	3.106	4.437	5.921
12	1.356	1.782	2.179	2.681	3.055	4.318	5.694
13	1.35	1.771	2.160	2.650	3.012	4.221	5.513
14	1.345	1.761	2.145	2.624	2.977	4.140	5.363
15	1.341	1.753	2.131	2.602	2.947	4.073	5.239
16	1.337	1.746	2.120	2.583	2.921	4.015	5.134
17	1.333	1.740	2.110	2.567	2.898	3.965	5.044
18	1.33	1.734	2.101	2.552	2.878	3.922	4.966
19	1.328	1.729	2.093	2.539	2.861	3.883	4.897
20	1.325	1.725	2.086	2.528	2.845	3.850	4.837
21	1.323	1.721	2.080	2.518	2.831	3.819	4.784
22	1.321	1.717	2.074	2.508	2.819	3.792	4.736
23	1.319	1.714	2.069	2.500	2.807	3.767	4.693
24	1.318	1.711	2.064	2.492	2.797	3.745	4.654
25	1.316	1.708	2.060	2.485	2.787	3.725	4.619
26	1.315	1.706	2.056	2.479	2.779	3.707	4.587
27	1.314	1.703	2.052	2.473	2.771	3.690	4.558
28	1.313	1.701	2.048	2.467	2.763	3.674	4.530
29	1.311	1.699	2.045	2.462	2.756	3.659	4.506
30	1.31	1.697	2.042	2.457	2.750	3.646	4.482
40	1.303	1.684	2.021	2.423	2.704	3.551	4.321
60	1.296	1.671	2.000	2.390	2.660	3.460	4.169
100	1.292	1.660	1.984	2.364	2.626	3.390	4.053
∞	1.282	1.645	1.960	2.326	2.576	3.291	3.750

Critical Value Table: t-table used for finding z* for a certain level of confidence.

Fig. (13). Critical values of the t distribution [47].

Estimating a Population Proportion

It is useful to depend on the proportions seen within a population sample to estimate a population proportion of an attribute.

You don't have to be a math major or a skilled statistician to understand the following concepts:

It would be beneficial to depend on the proportions seen within a population sample to estimate some characteristics within a population.

If you're going to depend on a sample, be sure it's a random sample. That indicates that the sample was drawn so that every member of the underlying population had an equal probability of being chosen.

It is critical to consider the sample size. The "confidence" that the observed sample percentage will be "near" to the actual population proportion grows as the size of a random sample grows. It's not unusual if you throw a fair coin 10 times and only receive three or fewer heads (a sample proportion of 30 percent or less). However, if there were 1,000 tosses, most individuals would agree that only 300 or fewer heads would be very implausible based on intuition and common experience. In other words, the greater the sample size, the more likely the sample percentage will be closer to the real "population" proportion of 50% [47].

While the sample percentage may be the most accurate estimate of the whole population proportion, you cannot be certain that it is the same proportion as the entire population proportion [47].

Using Confidence Intervals to Determine the Population Proportion

Consider the following scenario. Assume a political pollster selects 400 people and finds 208 for Candidate A and 192 for Candidate B. It leads to an estimate of 52 percent of A's support in the public. However, it is improbable that A will get precisely 52 percent of the vote. We'll refer to it as 0.52 p^ (pronounced "p-hat"). The sample proportion p is used to estimate the population proportion, p^. The estimate, however, is normally wrong by a factor known as the standard error (SE). The SE may be determined using the following formula:

$$\sqrt{\frac{\hat{p}\,(1-\hat{p})}{n}}$$

Where n is the number of people in the sample; as a result, the SE in this example is around 0.02498. As a result, for this example, a desirable population proportion would be: 0.52±0.2498 [47].

Statistical experts often employ particular confidence ranges for p. It is calculated in a slightly different way, using the following formula:

$$\hat{p} \pm z * \sqrt{\frac{\hat{p}\,(1 - \hat{p})}{n}}$$

Where z* is the standard normal distribution's upper critical value, if we wanted to calculate p with a 95% confidence level, we would use a Z -value of 1.960 (found using a critical value table) and find p to be 0.52± 0.04896. So, with 95 percent confidence, we can predict that between 47.104 percent and 56.896 percent of voters will vote for candidate A [47].

A basic rule of thumb is that if you employ a confidence level of X%, you should anticipate (100 X)% of your findings to be erroneous. If you select a 95 percent confidence level, you should anticipate that 5% of your findings will be erroneous [47].

A phenomenon that occurs often during an electoral year is the publication of news pieces that include confidence interval estimates in terms of percentages or proportions [47]. For example, a poll performed for a presidential candidate may indicate that the candidate received 40 percent of the vote and was in the range of three percentage digits of winning. The true proportion of voters who backed the candidate will be between 0.37 and 0.43 when election polls are estimated with 95 percent confidence: (0.40 – 0.03, 0.40 + 0.03) [47].

Investors are interested in the real percentage of equities that rise and fall in value each week when it comes to the stock market. Small businesses that sell personal computers are particularly interested in the percentage of homes in the United States who possess a computer. In order to determine the genuine percentage of stocks that move up or down each week and the true proportion of homes in the United States that possess personal computers, confidence intervals must first be established for both variables [47].

The technique for determining the confidence interval, the sample size, the error limit, and the confidence level for a percentage is similar to the procedure for determining the population mean, but the formulae used to calculate these values are different [47].

What are the signs of dealing with an issue of proportion?

To begin, we'll look at the binominal distribution, which is the most frequent kind of distribution. B (p, n) is the binomial coefficient for every binomial variable, where n the total number of tries and p is the probability that the trial will be successful on the first try. To find out what percentage of trials were successful, divide X, that represented the number of successful trials, by n (the total number of trials) (or the sample size) [53].

That percentage is represented by the random variable P′ (also known as "P prime").

$$P' = X / n.$$

When n is big and p is not near to zero or one, the normal distribution may be used to approximate the binomial distribution, and *vice versa*.

$$X \sim N\left(np, \sqrt{npq}\right.$$

A normal distribution of proportions with P′ as the random variable and the mean and standard deviation as the means and standard deviations is obtained by dividing the random variable by n and the mean and standard deviation by n. (Recall that a percentage is defined as the number of successes divided by the number of failures).

$$\frac{X}{n} = P' \sim N\left(\frac{np}{n}, \frac{\sqrt{npq}}{n}\right)$$

Using algebra to elaborate

$$\frac{\sqrt{npq}}{n} = \sqrt{\frac{pq}{n}}$$

When it comes to proportions, P′ follows a normal distribution:

$$\frac{X}{n} = P' \sim N\left(\frac{np}{n}, \frac{\sqrt{npq}}{n}\right)$$

It is only possible to employ a confidence interval if the number of successes np′ and the number of failures nq′ are both more than five. It is comparable to the error bound formula for a mean, except that the "acceptable standard deviation" is calculated differently [53].

When we know the population's standard deviation, we can calculate the proper standard deviation to apply for a given mean [53].

$$\sigma/\sqrt{n}.$$

In the case of a percentage, the relevant standard deviation is

$$\sqrt{\frac{pq}{n}}.$$

However in error bound equation we use [53]

$$\sqrt{\frac{p\prime q\prime}{n}}.$$

And for Standard Deviation (SD)

$$\sqrt{\frac{pq}{n}}.$$

It should be noted that the sample proportions p′ and Q′ are approximations of the unknown population proportions p and q in the error bound calculation. Because p and q are unknown, the approximated proportions p′ and q′ are used in place of them. Calculated from the data, the sample proportions p′ and q′ are as follows: p′ denotes the estimated percentage of successes, while q′ indicates the estimated proportion of failures [53].

Examples

Example 1

Suppose that a market research firm is hired to estimate the percent of adults living in a large city who have cell phones. Five hundred randomly selected adult residents in this city are surveyed to determine whether they have cell phones. Of the 500 people surveyed, 421 responded yes – they own cell phones. Using a 95% confidence level, compute a confidence interval estimate for the true proportion of adult residents of this city who have cell phones [53].

Calculate a confidence interval estimate for the real percentage of adult inhabitants of this city who own mobile phones based on a 95 percent confidence level.

Solution

Let X represent the number of persons in the sample who are in possession of a mobile phone. You must first determine p′, q′, and EBP in order to construct the confidence interval:

$n = 500$

x = the number of successes = 421

P'= xn

=421500

= 0.842P'

has been calculated as a percentage of a sample; this is a point estimate of the population proportion. $q' = 1 - p' = 1 - 0.842 = 0.158$

Given that CL = 0.95, 0.05 is equal to $1 - CL = 1 - 0.95 = 0.05$.

$\sigma/2 = 0.025$

Then $Z_{\sigma/2} = Z_{0.025} = 1.96$

Use the TI-83, 83+, or 84+ calculator command InvNorm(0.975,0,1) to find $Z_{0.025.}$ [53]

Keep in mind that the area to the right of $Z_{0.025}$ is 0.025 and the region to the left of $Z_{0.025}$ is 0.975 of the total area.

Additionally, this may be determined by using the proper instructions on other calculators, by using a computer, or by utilizing a Standard Normal probability table.

EBP = $(Z_{\sigma/2})$ (p'q'n) = (1.96) (0.842)(0.158)500 = 0.032

p' − EBP = 0.842 − 0.032 = 0.81

p' + EBP = 0.842 + 0.032 = 0.874

The real binomial population proportion has a 95 percent confidence interval of $(p' - EBP, p' + EBP) = (p' - EBP, p' + EBP)$ (0.810, 0.874) [53].

The "Plus Four" Confidence Interval

When computing a confidence interval for a percentage, there is little error injected into the process. We are compelled to utilize point estimates to determine the proper standard deviation of the sampling distribution since we do not know the real population percentage. Studies have demonstrated that the standard deviation estimate obtained as a consequence of this process might be inaccurate [53].

Fortunately, we can make a simple change that will enable us to construct more precise confidence intervals. For the sake of argument, let us imagine four more observations. Two of these observations were successful, while the other two were unsuccessful. As a result, the new sample size is $n + 4$, and the new number of successes is $x + 2$ [8].

The efficiency of this strategy has been established by computer research. This method should be used whenever the necessary confidence level is at least 90 percent, and the sample size is at least 10 [53].

Calculating the Sample Size n

If researchers want to be certain that their results will have a specified margin of error, they may use the error bound formula to determine the sample size that will be needed. The error bound formula for a population percentage is given by the following:

$$EBP = (Z_{\sigma/2}) \left(\sqrt{\frac{p'q'}{n}} \right)$$

When you solve for n, you will get an expression for the sample size [53].

$$n = (Z_{\sigma/2})^2 \, (p'q') \, / \, EBP^2$$

Sample Size Considerations

When determining the appropriate sample size, keep the following factors in mind:

When determining the appropriate sample size, we must take into account the following factors:

- What are the parameters of the population that we wish to estimate?
- The expense of sampling (importance of information)
- What is already known about the situation
- The population's spread (variability) is measured in percentages.
- Practicality: How difficult is it to get information?
- What level of precision do we want in the final estimations [8]

The Cost of Collecting Samples

In order to evaluate how exact our predictions should be, we need to consider the problem of sample costs. As we shall see later, while determining sample sizes, we must consider risk values and sample sizes. If the judgments we will make as a result of the sampling activity are really important, then we will seek low risk values and, as a result, higher sample sizes for the sampling activity [53].

Pre-existing knowledge

If our process has been researched before, we may utilize that past understanding to minimize the number of samples we need to collect. By using previous mean and variance estimations and stratifying the population to limit variation within groups, this may be accomplished [8].

Variability That is Inherent

When we collect samples, we may use them to estimate some attributes of the population that we are interested in. The variance of the estimate is equal to the inherent variability of the population divided by the number of participants in the study [53]:

$$\mathrm{Var}(p^{\wedge}) \approx \sigma^2/n.$$

in where p is the parameter that we are attempting to estimate This indicates that if the variability of the population is high, we will need to collect a large number of samples. In contrast, a low level of population variation indicates that we don't need to collect as many samples [53].

Determination of the Sample Size

Briefly stated, the stages involved in determining sample size are as follows:

It is necessary to include a statement describing the sample's anticipated. Before we can estimate anything, we must first establish what we are attempting to estimate, how exactly we want the estimate to be, and what we intend to do with the estimate once we have it. This should be straightforwardly deduced from the objectives [53].

We must devise an equation that relates the required accuracy of the estimate with the sample size in order to get the necessary precision. This is a remark about the likelihood of something happening. A handful of examples are provided below; consult your statistician if they are not relevant for your scenario [53].

There may be unknown characteristics of the population in this equation, such as the mean or variance, which must be calculated. Prior knowledge may be quite beneficial in this situation [53].

If you are stratifying the population in order to decrease variance, you must do sample size determination for each stratum separately [53].

The final sample size should be examined for its feasibility before being used. The only method to minimize it is to accept a less accurate sample estimate if it is deemed too high.

Proportions of Samples Taken

The probability statement regarding the required accuracy is the starting point for sampling proportions when doing sampling proportions calculations. This information is provided by [53]:

$$Pr(|p^\wedge - P| \geq \delta) = \alpha$$

- $P^\wedge$ represents the estimated percentage (p).
- P is the population parameter that is unknown.

The specified accuracy of the estimate is denoted by the symbol

The probability value is denoted by the symbol (usually low)

This equation simply demonstrates that we want the likelihood that the accuracy of our estimate will be less than we desire to be equal to a certain value. Of course, we like to keep the.1 value as low as possible. By making certain assumptions about the percentage being generally normally distributed, we can get an estimate of the needed sample size, which looks like this [53]:

$$n = z^2\alpha \ (pq \ / \ \delta^2)$$

Where z is the variate on the Normal curve that corresponds to the value of α.

CONCLUSION

Confidence intervals and sample sizes are important tools for researchers in statistics to be used to determine the population proportion. There are several tools that can be used to correctly construct these intervals as well as to estimate the target parameters. All are useful and important for understanding and estimating correctly.

CHAPTER 8

Hypothesis Testing

Abstract: Hypothesis testing is a simple way to construct information about a dataset and to determine if that information is true or false. Another method to think about hypothesis testing is to use it as a way to decide whether an assumption about a dataset succeeds and is supported or fails and is not supported. It is used frequently in science and medicine to better understand the effects of drugs and treatments.

Keywords: Alternative, Hypothesis, Null.

INTRODUCTION

Hypothesis testing evaluates two statements about a population to determine which is supported by the data. The two hypotheses are called the null hypothesis and the alternative hypothesis [54]. The null hypothesis states that there is no effect or the effect is equal to zero, while the alternative hypothesis states that there is an effect or the effect is not equal to zero. Hypothesis testing is used frequently in both the medical industry as well as in scientific research to determine if the results of an experiment within a sample population are statistically significant or have an effect [54]. It is important to understand whether the desired outcome of an experiment is for the null hypothesis to be true or for the alternative hypothesis to be true. For example, if researchers were testing a new drug for harmful side effects, they would want the null hypothesis for their results to be true, or that the side effects would be equal to zero. If the researchers were instead curious about whether the drug had an effect on the disease they were treating, they would want the alternative hypothesis to be true, or the drug to have an effect on the disease. Hypothesis tests use a random sample to draw conclusions about entire populations, so they are not 100% accurate [55]. There are two types of errors that can occur in hypothesis testing; false positives and false negatives [55]. False positives occur when the null hypothesis is rejected, but it is true, while for false negatives, the null hypothesis should be rejected and is not. Hypothesis tests rely on measurements of significance to determine how strongly the sample results must contradict the null hypothesis. The measurement of significance is called alpha and is set before the study begins. It can be thought of as the probability that the research says there is an effect when there is no effect. Lower levels of significance require stronger evidence to be true. For example, a 0.05 level of significance states that there is a 5% chance of deciding an effect exists when there is not an effect, while a 0.01 level of significance lowers this threshold to a 1% chance. Significance levels can also be visualized as critical regions on a normal curve. For example, a 0.05 significance level would result in shading in the far ends of the normal curve, 0.025 from each end. Both ends are used when it is a two-tailed hypothesis test, indicating that there could be potentially positive or negative effects shown in the test [56]. The term tail refers to the ends of the normal distribution curve. One-tailed hypothesis tests only test for effects in one direction, either positive effects above the mean or negative effects below the mean. Two-tailed tests are used when researchers are curious about any effect in a population or sample, while one-tailed tests are used either when researchers only care about an effect in one direction, or the effect can only occur in one direction [56]. These classic methods are provided to the reader as an introduction to the subject. There are other methods, which the reader is invited to seek out and explore on their own, should they find the subject interesting.

Z-Test

The Z-test is a type of statistical test that is used to determine if two population means are different when their variances are known, and the sample size is large [57]. Z-scores are the resultant of the test and are used in hypothesis testing. When a z-score is used, the data also follows a normal distribution. The standard deviation should also be known when a z-test is to be performed. In order for a Z-test to be conducted, the sample should also be randomized and all of the sample values independent from one another [57]. The Z-score derives from how many standard deviations above or below the mean the Z-test result is. Types of tests that can be performed using a Z-test include one and two-sample tests, paired difference tests and a maximum likelihood estimate.

Z -tests take a few steps to calculate: the alternative and null hypothesis should be stated, the alpha level and critical value of z should be found, and then the result of the z-test is compared to the critical z value to determine if the null hypothesis should be supported or rejected [57].

A one-sample Z-test is performed to compare the sample mean with the population mean. In order to complete a one-sample z-test, the following formula is used [57]:

$$\text{z score} = \frac{\bar{x} - \mu}{\sigma / \sqrt{n}}$$

where x bar represents the sample mean, mu represents the population mean, alpha is the population standard deviation, and n is the number of samples. If the calculated Z-score is greater than the critical value, then the null hypothesis can be rejected, meaning that there is an effect or the effect of the experiment is not zero.

A two-sample Z-test is used to compare the mean of two samples. It is represented by the following formula, where 1 and 2 are used to represent each of the samples [57]:

$$z \text{ score} = \frac{(\bar{x}_1 - \bar{x}_2) - (\mu_1 - \mu_2)}{\sqrt{\frac{\sigma_1^2}{n_1} + \frac{\sigma_2^2}{n_2}}}$$

the x bar calculation is used to represent the differences between the sample means, the mu calculation represents the difference between the population means, each alpha reflects the population standard deviation in each set, and each n represents the number of samples in each set. As before, if the Z-score result is greater than the critical value, the null hypothesis is rejected [57].

A paired difference Z-test is used to determine if the mean difference between two populations is greater than, less than or equal to zero. For a paired sample Z-test, the usual Z-test population requirements of a large data set that is normally distributed and randomly sampled apply as well as two additional requirements that the data must be continuous and the two sets, should show a similar spread between groups. Two calculations are made for a paired different Z-test. First, the paired differences in each sample are calculated using the following equation [57]:

$$\bar{X} = \frac{\sum_{i=1}^{n} X_i}{n}.$$

where Xi represents each sample up to the total number of samples present, and n is the total number of sample. This X bar value is then input into the following formula to find the Z-score [57]:

$$z = \frac{\bar{X}}{\sigma/\sqrt{n}}$$

where alpha is the standard deviation, as we have seen before. The result is then compared to the p-value to reject or accept the hypothesis.

A maximum likelihood estimate can also be used as a Z-score value to accept or reject the null hypothesis. In this usage, the maximum likelihood estimate divided by its standard error is used as the z-test statistic. The following formula is used in this instance [57]:

$$(\hat{\theta} - \theta_0)/\mathrm{SE}(\hat{\theta})$$

where theta is the maximum likelihood, theta zero is the value of theta under the null hypothesis and SE theta is the standard error associated with the maximum likelihood.

T-Test

The t-test or Student's t- test is a common statistical test that is used to determine if there is a significant difference between the means of two groups [58]. It is so named as it first appeared in a paper in the scientific journal Biometrika, where the author, William Sealy Gosset, used the pseudonym Student [58]. Gosset worked at the Guinness brewery in Ireland and was interested in the properties of barley in small sample sizes. He devised the test to monitor the batch quality of different versions of the Guinness stout beer.

There are both one and two sample versions of the t-test. The one-sample version is used to determine whether the mean of a population has a value specified in the null hypothesis. The two-sample version, also known as the Student's t-test, is used to determine if the means of two populations are equal, when the variances of the two populations are also assumed to be equal [59]. When this assumption is dropped, the test is referred to as Welch's t-test.

The formula used for a one-sample t-test is as follows [59]:

$$t = \frac{Z}{s} = \frac{\bar{X} - \mu}{\hat{\sigma}/\sqrt{n}}$$

where x bar is the sample mean, mu is the population mean, alpha is the variance, and n is the number of samples. S represents the standard error of the mean where the sample variance follows a scaled X2 distribution. This equation assumes that the population follows a normal distribution and that Z and s are independent variables.

For a two-sample t-test, there is additional complexity. For the two sample tests, it is assumed that the means of the two populations follow normal distributions, they have the same variance and that the sampled data from each population be independent. Two sample tests can involve unpaired or paired samples. For unpaired samples, two sets of independent and identically distributed samples are used, one from each of the two populations that are being compared [60]. An example of an unpaired sample test would be to compare the effects of a drug on a test group *versus* a control group.

The formula for this kind of t-test is as follows $t = \dfrac{\bar{X}_1 - \bar{X}_2}{s_p \sqrt{\frac{2}{n}}}$ [59]:

where [59]

$$s_p = \sqrt{\frac{s_{X_1}^2 + s_{X_2}^2}{2}}.$$

In this formula, sp is the pooled standard deviation for $n = n_1 = n_2$ and s^2_{x1} and s^2_{x2} are the unbiased estimators of the variances of the two samples. The denominator of t is the standard error of the difference between two means. This formula is only used when the two samples are of the same size and have equal variance [60]. If the variance is not equal, then Welch's t-test is used and has the following form [60]:

$$t = \frac{\bar{X}_1 - \bar{X}_2}{s_{\bar{\Delta}}}$$

where $s_{\bar{\Delta}} = \sqrt{\dfrac{s_1^2}{n_1} + \dfrac{s_2^2}{n_2}}.$ [60]

where si2 is the unbiased estimator of the variance of each of the two samples with ni = number of participants in group i (1 or 2). In this case, s delta is not a pooled variance.

For a paired sample t-test, the samples would consist of matched pairs of units or of one sample that has been tested twice. An example of a paired test would be comparing the before and after effects of a group on the same group; essentially, this would mean that each individual would be their own control [60]. This improves the potential of rejection of the null hypothesis that there is no effect as it eliminates interpatient variance as a confounding factor in the analysis. The

formula for the dependent t-test is as follows $t = \dfrac{\bar{X}_D - \mu_0}{s_D / \sqrt{n}}$ [60]:

where XD and sD are the average and standard deviation of the differences between all pairs. The pairs are *e.g.,* either one person's pre-test and post-test scores or between-pairs of persons matched into meaningful groups (for instance, drawn from the same family or age group. Mu zero is equal to the value of zero if we want to test if the average of the difference is significantly different. This equation uses n-1 degrees of freedom where n the number of pairs.

CONCLUSION

To conclude, the importance of hypothesis testing cannot be understated. It is an extremely valuable tool to understand the integrity of a dataset. By using hypothesis testing, such as t-tests and z-tests, researchers can elucidate whether information or assumptions within a dataset are true or false. By being able to determine whether an assumption holds up under testing, researchers can then decide how to analyze the dataset and if additional testing is needed. By using hypothesis testing, the direction of future work and its effects on a population can be better understood. Hypothesis testing is also valuable as a measure of the deviation between two groups of means and is most frequently used in science to help understand effects or trends present during the experiment. Without hypothesis testing, researchers would not have an effective way to measure changes in a population or to understand those changes in the context of existing information.

CHAPTER 9

Correlation and Regression

Abstract: Researchers use correlation and regressions to show the relationships between factors or scenarios in a dataset. These types of analyses are important to all aspects of statistical analysis and are a common way to report connections between factors in a data set. This section will address basic linear correlation and regression techniques.

Keywords: Correlation, Linear, Negative, Positive, Regression.

INTRODUCTION

Correlation and regression are complex and powerful techniques that are used in data analysis. Correlation and regression are used to analyze the relationship between two continuous variables. In general, the dependent or outcome variable is referred to as Y, and the independent or predictor variable is referred to as X [61]. This type of analysis is used in various disciplines, but is most common in science and technology, when researchers are trying to understand how aspects of the data affect one another and to make predictions about the future dataset that include those parameters.

CORRELATION

Correlation quantifies the direction and strength of the relationship between two or more numeric variables. It lies between +1.0 and -1.0. When the correlation is negative, the slope of the regression line will also be negative and vice versa [62]. The correlation squared, which is written as R^2, has a special meaning in simple linear regression. The R^2 value represents the proportion of variation in Y as explained by X. For correlation, X and Y data can be used interchangeably. Also, in correlation, X and Y are random variables. Correlation is a more precise summary of the relationship between two variables [63]. The result of a pairwise correlation can be gathered together in a table to summarize relationships within the data set. Variables are considered to be uncorrelated when a change in one variable does not result in a change in the other variable. When both variables move in the same direction, or an increase in X results in an increase in Y, this is considered a positive correlation [64]. For example, the demand and price of a product are often positively correlated. As the demand for the item increases, so does the price of the item. When the variables are moving in opposite directions, or an increase in X causes a decrease in Y, this is a negative correlation [62]. The example of price and demand can also work here, as an increase in the price of the product will result in a decrease in demand as less people are willing to pay the increased price. Correlations allow for the association or absence of a relationship between two variables [65]. If we can show two variables are associated or correlated, then the strength of the relationship between the two can be measured. This allows for predictions using one factor to be applied to additional factors as we know how a change in one aspect will positively or negatively affect the other and by what degree, based on the direction and strength of the correlation [62]. Correlations are visualized using scatter plots, as shown in Fig. (**1**).

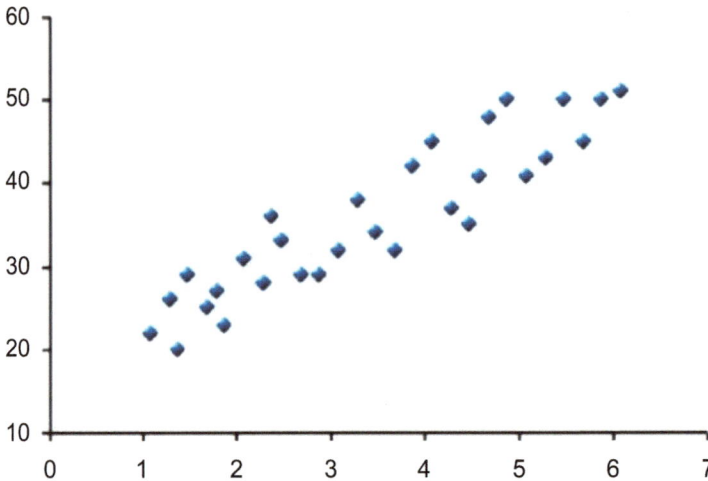

Fig. (1). Correlation scatters plot. [Source: Wikimedia Commons, By Jsmura - Own work, CC BY-SA 4.0].

REGRESSION

Regression is the relationship between the independent and dependent variables to each other. In simple linear regression, this relationship is the relationship of X to Y and is described as the equation of the line $Y = a + bX$. Regression attempts to explain how X causes Y to change and will generate different results if X and Y are reversed [62]. Regression assumes that X is a fixed value with no error. As regression analysis generates a linear equation, this analysis can be used for prediction or optimization and used to make assumptions about similar data sets. Regression can be described as how one variable affects another, or how changes in a variable trigger change in another variable, essentially cause and effect [66]. We can use a simple example of agriculture to understand regression. If there is ample rainfall, then seeds planted in a field will grow. If there is a drought and no rain falls, then the seeds will not grow. Regression analysis is used to determine the functional relationship between two variables, X and Y to make predictions about an unknown variable Z [62]. The value of this unknown variable Z can be estimated based on the values of the fixed variables X and Y. For simple linear regression, the best fit of the line through the data points is used to generate an equation that can then be used with the unknown Z variable [67]. A regression plot takes the correlation plot one step further by fitting a line through the scattered data points. An example regression is shown in Fig. (**2**).

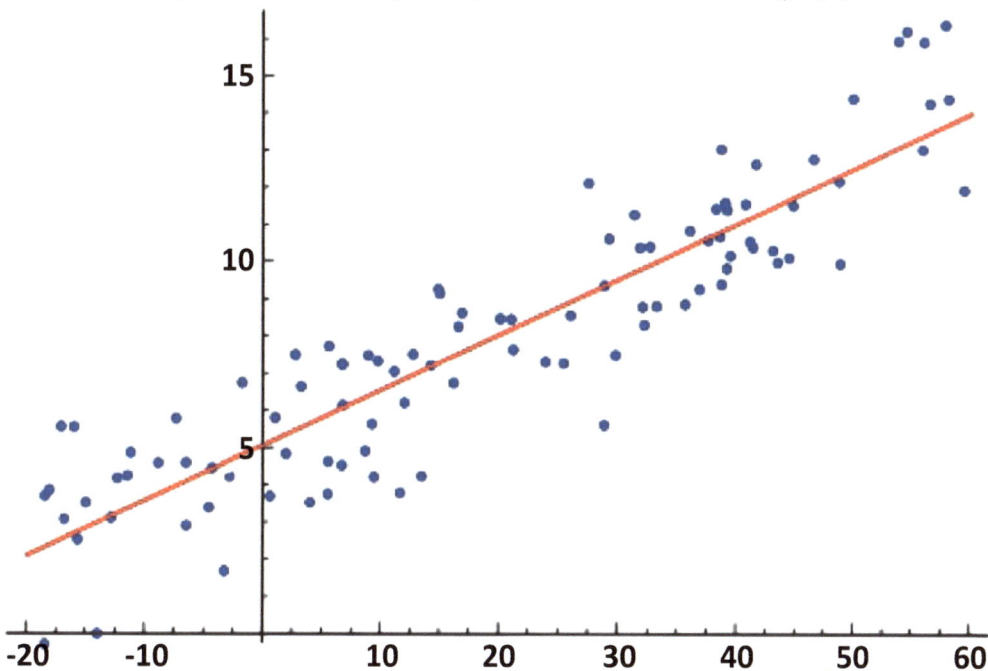

Fig. (2). Random data points and their linear regression. [Source: Wikimedia Commons Created with the following Sage (http://sagemath.org) commands: X = RealDistribution('uniform', [-20, 60]) Y = RealDistribution('gaussian', 1.5) f(x) = 3*x/20 + 5 xvals].

CONCLUSION

In conclusion, linear regression and correlation are one of the simplest and most common ways to show the relationship between independent and dependent variables to one another. It is a very powerful tool in statistics. Linear regressions help researchers to visualize these relationships and thereby assist in understanding the strength of similarities or differences within a dataset. By using these tools, researchers are able to better elucidate the outcomes of the inclusion of new variables as well as the relationships between relevant datasets.

CHAPTER 10

Ethics

Abstract: Ethical behavior by researchers, analysts and statisticians is paramount to the creation and reporting of data in all fields. Without ethics, data obtained and analyzed cannot be relied upon. A short discussion of ethical standards for all involved in data handling, research and statistical analysis is included, as this important aspect of the field cannot be understated.

Keywords: Ethics, Integrity, Standards.

INTRODUCTION

An overview of common statistics would be incomplete without the inclusion of a discussion on ethics within the field. The field of statistics must also include ethics as a way to govern both the integrity of researchers and the data set that those researchers analyze and handle. Grounding in ethics allows for confidence in reported results as well as integrity for future works and analysis of those experiments and investigations. The ethical standards associated with the field of statistics are laid out by the American Statistical Association.

ETHICS

Statistics is the science of uncertainty and variation, but the paradox is that in scientific research, statistics are used to justify claims or solidify determinations. Statistics are meant to learn from data, and gather insights about future trends and plan further investigations based on the current datasets. There are necessary considerations statisticians must wrestle with in order to effectively communicate the uncertainty and variation in their analyses [68]. To assist researchers with these considerations, the American Statistical Association has set out eight guidelines for ethical practice. First, the guideline of professional integrity and accountability. For statisticians to adhere to this guideline, they agree to use the methodology that is relevant and appropriate to produce valid, interpretable and reproducible results [69]. The statistician also agrees to accept responsibility for their professional work, credit others whose work is used and not perform work beyond their scope of expertise [71]. The second guideline follows from the first and specifies that the statistician is candid and upfront about any deficiencies, limitations or biases in the data [72]. This includes acknowledgement of assumptions, outlining methods to ensure the integrity of the dataset, and addressing potential confounding variables that are not present in the study [69]. The ethical statistician also has a responsibility to the funder, client and scientific community as outlined in the third guideline [70]. This means that the statistician strives to make new knowledge available to the scientific community, applies analyses scientifically, and protects the use and disclose of data appropriately [69]. The fourth guideline of responsibility to the research subjects follows directly from the previous guideline. As part of this guideline, the statistician protects and preserves the rights and data of all studied subjects to the best of their ability as well as providing participants with all appropriate disclosures, releases and research approvals [69]. The fifth guideline addresses the responsibility of the statistician to their research colleagues by promoting transparency in study design and analysis as well as effectively communicating and reviewing findings with other researchers [69]. The sixth guideline addresses the responsibility to other statisticians or statistics practitioners. Out of respect for other statisticians, the leader of the study agrees to treat others with respect and focus on scientific principles, methodology and the substance of data interpretations [69]. The seventh guideline notes the responsibility of the statistician regarding scientific misconduct. The ethical statistician avoids misconduct and does not condone questionable practices with regard to scientific, professional or statistical misconduct [69]. The final guideline for ethical statisticians addresses the responsibility of employers to statisticians. This guideline says that those employing statisticians agree to respect their professional expertise, support sound analysis and promote a safe and ethical working environment [71]. By adhering to these eight guidelines, statisticians ensure that they are acting in an ethical manner and best representing the science of statisticians and their professional colleagues in the discipline.

CONCLUSION

Ethics in statistical analysis are paramount. Researchers and statisticians must show integrity in their work and analyses. It is the gravest responsibility of the person doing statistics to upload the ethical standards and practices of the profession as well as to address any potential conflicts of interest or biases within their analysis. Upholding ethical standards is important for the public trust of data analysts with respect to their professional reputation of themselves as well as their colleagues.

REFERENCES

[1]   D. Montgomery, G. Runger, and N. Hubele, "Engineering Statistics, 5th Edition", 2010. Available at:https://www.wiley.com/en-us/Engineering+Statistics%2C+5th+Edition-p-9780470631478

[2]   Population *vs* Sample Data – Math Bits Notebook (A1 - CCSS Math)," Available at:, https://mathbitsnotebook.com/Algebra1/StatisticsData/STPopSample.html

[3]   Sample Dataset - an overview | ScienceDirect Topics, Available at:, https://www.sciencedirect.com/topics/computer-science/sample-dataset

[4]   L. Daniels, and N. Minot, "An Introduction to statistics and data analysis using stata®: from research design to final report", *Sage Publications 2019.*

[5]   R.A. Johnson, and G.K. Bhattacharyya, "Statistics: principles and methods", *John Wiley & Sons, 2019.*

[6]   J.E. Kolassa, "An introduction to nonparametric statistics", *Chapman and Hall/CRC, 2020.*

[7]   C. Chatfield, "Statistics for technology: a course in applied statistics", *Routledge, 2018.*

[8]   S.M. Ross, "Introduction to probability and statistics for engineers and scientists", *Academic press 2020.*

[9]   S. Manikandan, "Frequency distribution", *J. Pharmacol. Pharmacother. 2011,* vol. 2, no. 1, pp. 54-56.

[10] "Pie Charts, Histograms, and Other Graphs Used in Statistics, Available at: https://www.thoughtco.com/frequently-used-statistics-graphs-4158380

[11] D. Russell, *What Is a Bar Graph?*. https://www.thoughtco.com/definition-of-bar-graph-231238

[12] "A Complete Guide to Pie Charts,", https://chartio.com/learn/charts/pie-chart-complete-guide/

[13] 13.2 - Stem-and-Leaf Plots | STAT 414, Available at: https://online.stat.psu.edu/stat414/lesson/13/13.2

[14] Statistics: Power from Data! Organizing data: Stem and leaf plots, Available at: https://www150.statcan.gc.ca/n1/edu/power-pouvoir/ch8/5214816-eng.htm

[15] "IXL | Dot plots," Available at:, https://www.ixl.com/math/lessons/dot-plots

[16] S. Foster, "The Federal Reserve's Dot Plot Explained – And What It Says About Interest Rates," Available at:, https://www.bankrate.com/banking/federal-reserve/federal-reserve-dot-plot-explained-how-to-read-interest-rates/

[17] "A Complete Guide to Scatter Plots," Available at:, https://chartio.com/learn/charts/what-is-a-scatter-plot/

[18] "Create and use a time series graph—ArcGIS Insights | Documentation," Available at:, https://doc.arcgis.com/en/insights/latest/create/time-series.htm

[19] E.G.E. Kyonka, S.H. Mitchell, and L.A. Bizo, "Beyond inference by eye: Statistical and graphing practices in JEAB, 1992-2017", *J. Exp. Anal. Behav.,* vol. 111, no. 2, pp. 155-165. *2019.*

[20] N. Barnett, "Graphing causation: getting a clearer picture or fuzzy logic? Comment on Br J Anaesth 2020: 125: 393–97", *Br. J. Anaesth.,* vol. 126, no. 3, pp. e100-e101.

[21] D.M. Finkelstein, and D.A. Schoenfeld, "Graphing the win ratio and its components over time", *Stat. Med. 2019,* vol. 38, no. 1, pp. 53-61.

[22] Ungrouped Data, "Introduction to Statistics 2019",

[23] C.E.L. Kinney, J.C. Begeny, S.A. Stage, S. Patterson, A. Johnson, "Three alternatives for graphing behavioral data: A comparison of usability and acceptability", Behavior Modification 46.1, 2022: 3-35.

[24]   M. Malloy, J. Koller, and A. Cahn, "Graphing crumbling cookies", *ACM, 2019.*

[25]   J.O. Aldrich, "Using IBM SPSS statistics: An interactive hands-on approach", *Sage Publications 2018.*

[26]   L. Sullivan, *The role of probability.,* .https://sphweb.bumc.bu.edu

[27]   F.V. Kuhlmann, *A simple explanation of probability. Licensed under Creative Commons by-nc-sa, 2021.*

[28]   A. Schmitz, *Basic Concepts of Probability. Licensed under Creative Commons by-nc-sa 3.0.* 2013, Available at: https://2012books.lardbucket.org/books/beginning-statistics/s07-basic-concept--of-probability.html

[29]   An Introduction to Probability and Statistics (Wiley Series in Probability and Statistics). New York, 2021.

[30]   W. Kenton, *"Random Variable." Investopedia: published by Dot dash Meredith Group. 2022.* Available at: https://www.investopedia.com/terms/r/random-variable.asp

[31]   A. Schmitz, *"Discrete Random Variables." Saylor Academy Texts, 2021. Available at:*.https://saylordotorg.github.io/text_introductory-statistics/s08-discrete-random-variables.html

[32]   J.M. Russell, *"Introduction to Discrete Random Variables and Notation." Pressbooks, Licensed under Creative Commons cc-by-sa 4.0.* 2022.

[33]   A. Barone, *"Binomial Distribution." Investopedia: published by Dot dash Meredith Group. 2021.* Available at: https://www.investopedia.com/terms/b/binomialdistribution.asp

[34]   C.F.I. Team, *Binomial Distribution. 2022.* https://corporatefinanceinstitute.com/resources/knowledge/ other/ binomial-distribution/

[35]   H.B. Berman, *Negative Binomial Distribution.* 2022, Available at: https://stattrek.com/probability-distributions/negative-binomial

[36]   M. Taboga, *"Continuous random variable", Lectures on probability theory and mathematical statistics. Kindle Direct Publishing, 2021. Available at:*.https://www.statlect.com/glossary/continuous-random-variable

[37]   M. Taboga, Absolutely Continuous random variable", Lectures on probability theory and mathematical statistics. Kindle Direct Publishing. 2021, Available at: https://www.statlect.com/ glossary/absolutely-continuous-random-variable

[38]   Maple Tech International. Continuous Random Variable., Available at: https://www.math.net/ continuous-random-variable

[39]   J. Chen, Normal Distribution, Available at: https://www.investopedia.com/terms/ n/normaldistribution.asp

[40]   M. Taboga, "Central Limit Theorem", Lectures on probability theory and mathematical statistics. Kindle Direct Publishing. Online appendix, 2021, https://www.statlect.com/asymptotic-theory/centra--limit-theorem

[41]   P. Bhandari, The Standard Normal Distribution, Available at: https://www.scribbr.com/statistics/ standard-normal-distribution/#:~:text=The%20standard%20normal %20distribution%2C%20also, the%20mean%20each%20value%20lies

[42]   Saylor Academy. Areas of tails of distributions, Available at: https://saylordotorg.github.io/ text_introductory-statistics/s09-04-areas-of-tails-of-distribution.html

[43]   Corporate Finance Institute, "Sampling Distribution", Available at: https://corporatefinance institute.com/resources/ knowledge/other/sampling-distribution/

[44]   A. Schmitz, Beginning Statistics. Sampling Distributions, Available at: https://2012books.lardbucket.org/books/beginning-statistics/s10-sampling-distributions.html

[45] Penn State Open Education. Sampling Distribution of the Sample Mean, Available at: https://online.stat.psu.edu/stat500/lesson/4/4.1

[46] Lani, James. Statistics Solutions. "Estimation.", Available at: https://www.statisticssolutions.com/estimation/

[47] D. Lane, *Introduction to Estimation. September 17, 2013. Provided by: OpenStax CNX. Available at:*. https://cnx.org/contents/5530cbcc-820d-4f48-83c7-fe03ef5823be@4 Available at:, https://courses.lumenlearning.com/boundless-statistics/chapter/estimation/

[48] H.B. Berman, *Estimation in Statistics.* https://stattrek.com/estimation/estimation-in-statistics.aspx

[49] H.B. Berman, *What is a Confidence Interval.* https://stattrek.com/estimation/confidence-interval https://stattrek.com/estimation/confidence-interval.aspx?tutorial=AP

[50] G. Stephanie. "Estimator: Simple Definition and Examples" From StatisticsHowTo.com: Elementary Statistics for the rest of us!, Available at: https://www.statisticshowto.com/estimator/

[51] H.B. Berman, *What is the Standard Error?*.https://stattrek.com/estimation/standard-error.aspx?tutorial=AP

[52] H.B. Berman, *Margin of Error.* https://stattrek.com/estimation/margin-of-error.aspx?tutorial=AP

[53] S. Andy. Beginning Statistics. " Estimation.", Available at: https://2012books.lardbucket.org/books/beginning-statistics/s11-estimation.html

[54] Frost, Jim. Statistics by Jim. "One-Tailed and Two-Tailed Hypothesis Tests Explained,", Available at: https://statisticsbyjim.com/hypothesis-testing/

[55] Frost, Jim. Statistics by Jim. "Null hypothesis," Available at: , https://statisticsbyjim.com/?s=null+hypothesis

[56] Frost, Jim. Statistics by Jim. "Hypothesis Tests Explained," Available at:, https://statisticsbyjim.com/hypothesis-testing/

[57] J. Chen, Z-Test Definition: Its Uses in Statistics Simply Explained With Example, https://www.investopedia.com/terms/z/z-test.asp

[58] S. Mina. Analytics Vidhya. "Hypothesis Testing | Difference between Z-Test and T-Test" Available at:, https://www.analyticsvidhya.com/blog/2020/06/statistics-analytics-hypothesis-testing-z-test-t-test

[59] R.C. Sprinthall, *(2011). Basic Statistical Analysis (9th ed.). Pearson Education. "Z-test," Available at:* .https://en.wikipedia.org/wiki/Z-test

[60] Stats Test Team. "Paired Samples Z-Test," Available at:, https://www.statstest.com/paired-samples-z-test/

[61] https://www.g2.com/articles/correlation-vs-regression

[62] B. Gerstman, *Correlation and Regression.*https://www2.sjsu.edu/faculty/gerstman/StatPrimer/cont-cont.htm

[63] J.M. Bland, and D.G. Altman, "Statistical methods for assessing agreement between two methods of clinical measurement", *Lancet,* vol. i, pp. 307-310. 1986.

[64] V. Bewick, L. Cheek, and J. Ball, "Statistics review 7: Correlation and regression", *Crit. Care,* vol. 7, no. 6, pp. 451-459. 2003.[http://dx.doi.org/10.1186/cc2401]

[65] P. Vadapalli, *Correlation vs Regression: Difference Between Correlation and Regression.* https://www.upgrad.com/blog/correlation-vs-regression/

[66] K. Kozak, *Regression and Correlation..* https://www.coconino.edu/resources/files/pdfs/academics/sabbatical-reports/kate-kozak/chapter_10.pdf

[67] S. Crawford, "Circulation. 2006;114: 2083–2088", [http://dx.doi.org/10.1161/circulationaha.105.586495]

[68]   A. Gelman, "Ethics in statistical practice and communication: Five recommendations", *Significance,* vol. 15, no. 5, pp. 40-43. 2018.

[69]   "Ethical Guidelines for Statistical Practice," https://www.amstat.org/ASA/Your-Career/Ethica-
-Guidelines-for-Statistical-Practice.aspx

[70]   L.M. Lesser, and E. Nordenhaug, "Ethical Statistics and Statistical Ethics: Making an Interdisciplinary Module", *J. Stat. Educ.,* vol. 12, p. 3. 2004. [http://dx.doi.org/10.1080/10691898.2004.11910630]

[71]   J.S. Gardenier, "Making Statistical Ethics Work for You, of Course", *Amstat News,* no. 296, pp. 21-22. 2002.

[72]   L.M. Lesser, *2001 Proceedings of the American Statistical Association Section on Statistical Education [CD-ROM], Alexandria, VA: American Statistical Association, 2001.*

SUBJECT INDEX

www.ingramcontent.com/pod-product-compliance
Lightning Source LLC
Chambersburg PA
CBHW041706210326
41598CB00007B/548